SpringerBriefs in Systems Biology

More information about this series at http://www.springer.com/series/10426

Meidjie Ang

Metabolic Response of Slowly Absorbed Carbohydrates in Type 2 Diabetes Mellitus

 Springer

Meidjie Ang
Justus Liebig University Giessen
Giessen, Germany
E-mail: meidjie.ang@gmail.com

This work was carried out at the Clinical Research Unit of Medical Clinic and Policlinic 3 and accepted as doctoral thesis at the Faculty of Agricultural Sciences, Nutritional Sciences, and Environmental Management of the Justus Liebig University Giessen.

ISSN 2193-4746   ISSN 2193-4754   (electronic)
SpringerBriefs in Systems Biology
ISBN 978-3-319-27897-1   ISBN 978-3-319-27898-8   (eBook)
DOI 10.1007/978-3-319-27898-8

Library of Congress Control Number: 2015959481

Springer Cham Heidelberg New York Dordrecht London

Printed on acid-free paper

Springer International Publishing AG Switzerland is part of Springer Science+Business Media (www.springer.com)

To Ronny and my family

Preface

This book focuses on relevant topic in clinical nutrition research, particularly on the effects of slowly absorbed carbohydrates on postprandial glucose metabolism in type 2 diabetes. Slowly absorbed carbohydrates will cause gradual increases in blood glucose and insulin levels and may therefore be effective as part of a treatment strategy for glycemic control and reduction of cardiovascular complications in type 2 diabetes.

Two clinical studies with randomized, double-blinded, and crossover design are presented in detail in this book. The first study characterized metabolic pathway of the disaccharide slowly digested isomaltulose as opposed to rapidly digested sucrose using a double isotope technique, which combines a hyperinsulinemic-euglycemic clamp with an oral load of the disaccharides. This method enables assessment of glucose absorption, postprandial glucose turnover, and insulin sensitivity by using mathematical models, such as the one- and two-compartment models of glucose kinetics as well as the oral glucose minimal model of insulin action in type 2 diabetic patients. The second study assessed postprandial glucose and insulin responses after oral administration of isomaltulose alone or in combination with proteins in type 2 diabetic patients. The results of these studies provide novel insight on the effects of slowly absorbed carbohydrates on postprandial glucose homeostasis in type 2 diabetes.

The key features of this book are:

- It provides an up-to-date systematic review on the effects of slowly absorbed isomaltulose on postprandial glucose metabolism in humans, including healthy probands, impaired glucose-tolerant subjects, and both type 1 and type 2 diabetic patients
- It examines the metabolic response of slowly absorbed isomaltulose compared to rapidly absorbed sucrose comprehensively in type 2 diabetic patients
- It gives an extensive description of the assessment methods and a detailed calculation procedure of glucose kinetics and insulin action
- It highlights new evidence on the role of slowly absorbed carbohydrates for glycemic control in type 2 diabetes

This book is intended for both the general reader and the specialist, including nutritional scientists, academics or researchers interested in nutrition and clinical nutrition, endocrinologists, diet or diabetes advisors, nutritionists, public health scientists or healthcare providers who are interested in health promotion and prevention of diabetes diseases, and students searching for detail information on the application of stable isotopes in a clinical setting.

September 2015   Meidjie Ang

Acknowledgments

This book is a hard work of many years. It would not have been possible without the encouragement, support, and contribution from a lot of people. First and foremost, I thank my supervisor Prof. Dr. Thomas Linn for giving me the opportunity to work in the challenging area of glucose homeostasis in diabetes mellitus. I appreciate all his contributions of time, ideas, and fruitful discussion toward this book. He has showed me directly and indirectly how to manage clinical research studies and taught me about patience and persistence when I hit a bottleneck during the time period of pursuing my doctoral degree. His suggestions, attention, and prompt responses were a great deal to my work. I am also grateful to Prof. Dr. Michael Krawinkel for his time to review the draft of my book and give valuable inputs.

My sincere gratitude is extended to the past and present group members of the Clinical Research Unit (CRU), especially Sina Streichert and Jutta Sitte, for taking care of the patients and for their technical assistance in performing the studies. I also thank Jutta Schmidt for her support in measuring glucagon in blood samples and Doris Erb for the guidance through the laboratory during my early stay in the Medical Clinic and Policlinic 3. I am also indebted to all the people who were in charge of additional blood samples.

In regard to the stable isotope probes, I acknowledge the support of Dr. Christian Borsch from the Analytical Platform of Stable Isotope and Cell Biology (Director: Prof. Dr. Clemens Kunz) in analyzing the preliminary stable isotope derivatives in blood and breath samples. Additional analyses were performed at the laboratory of the Clinical Research Center of the University of Rochester. I would also like to thank the Numico Research for contributing with the amino acid data and Prof. Richard Bergman at the Cedars-Sinai Diabetes and Obesity Research Institute for providing the OOPSEG program. Further, I express my sincere appreciation to Dr. Manfred Hollenhorst for introducing SPSS to me many years ago.

My time at CRU was also enriched by the Giessen Graduate Centre for the Life Sciences. I thank my fellow graduate students, Neelam, Chunguang, and Balaji, for sharing their experiences and opinions with me. For this book I thank Ronny and my brothers Husen and Tonny for their time, interest, and helpful comments in reading and editing the draft.

I deeply send my gratitude to my family in Indonesia for all their love, encouragement, and endless support: my parents who raised me with love, provided me with a good education, and have been so caring and supportive to me all the time, my grandpa for those sweet memories and for teaching me so much in my childhood, and my sisters and brothers for always motivating and supporting me. To Ronny whose love, patience, and support have continuously backed me up to keep moving on, you give me the courage to face ups and downs in the process. Thank you so much.

Last but not least, I thank the patients for their contributions to the studies.

July 2014   Meidjie Ang

Contents

Abbreviations

1-CM	one-compartment model
2-CM	two-compartment model
ADA	American Diabetes Association
ANOVA	analysis of variance
ATP	adenosine triphosphate
AUC	area under the curve
BCAA	branched-chain amino acids
BMI	body mass index
EAA	essential amino acids
EASD	European Association for the Study of Diabetes
EGP	endogenous glucose production
ELISA	enzyme-linked immunosorbent assay
FDA	Food and Drug Administration
FFA	free fatty acids
G-6-PD	glucose-6-phosphate dehydrogenase
G-6-P	glucose-6-phosphate
GCMS	gas chromatography mass spectrometry
GC	gas chromatograph
GDM	gestational diabetes mellitus
GI	glycemic index
GINF	glucose infusion
GIP	glucose-dependent insulinotropic polypeptide
GLP-1	glucagon-like peptide-1
GLUT	glucose transporter
HbA_{1c}	glycated hemoglobin
HOMA-β	homeostasis model assessment of β-cell function
HOMA-IR	homeostasis model assessment of insulin resistance
HOMA	homeostasis model assessment
HPLC	high-performance liquid chromatography
iAUC	incremental area under the curve
IDF	International Diabetes Federation

IFG	impaired fasting glucose
IGT	impaired glucose tolerance
IRMS	isotope ratio mass spectrometry
ISO+C	isomaltulose combined with casein
ISO+WS	isomaltulose combined with whey/soy (ratio 1:1)
ISO	isomaltulose
IVGTT	intravenous glucose tolerance test
MGTT	meal glucose tolerance test
MODY	maturity-onset diabetes of the young
MS	mass spectrometer
m/z	mass-to-charge ratio
NAD	nicotinamide adenine dinucleotide
NSB	nonspecific binding
OGTT	oral glucose tolerance test
PDB	Pee Dee Belemnite
ppm	parts per million
R_aO	oral rates of appearance
R_aT	total rates of appearance
R_a	rate of appearance
R_dT	total rates of disappearance
R_d	rate of disappearance
RIA	radioimmunoassay
SGLT1	sodium-dependent glucose transporter 1
SGU	splanchnic glucose uptake
SIM	selected ion monitoring
S_I	insulin sensitivity index
SUC	sucrose
T1DM	type 1 diabetes mellitus
T2DM	type 2 diabetes mellitus
TAA	total amino acids
TC	total count
TMB	tetramethylbenzidine
TTR	tracer-to-tracee ratio
WHO	World Health Organization

Chapter 1
Introduction

The prevalence of diabetes mellitus is increasing worldwide. Recent global estimations indicate that the number of people with diabetes has more than doubled over nearly the past three decades to 347 million (Danaei et al. 2011). World Health Organization (WHO) predicts that diabetes will be the seventh leading cause of death in 2030 (WHO 2011a). Much of the increase in diabetes incidence has been attributed to population growth, aging, and lifestyle changes including excessive calorie intake as well as physical inactivity that lead to obesity. The most common form of diabetes is type 2 diabetes mellitus (T2DM), affecting around 90 % of diabetic patients worldwide (WHO 1999). T2DM often occurs in adulthood; however, it has become increasingly prevalent in obese children and adolescents (Reinehr 2013). This metabolic disease is characterized by chronic fasting and postprandial hyperglycemia, with the central defects being defective insulin secretion and diminished tissue insulin sensitivity (WHO 1999).

1.1 Problem Statement and Aims

Increasing evidence suggests that postprandial hyperglycemia plays an important role in the development of long-term cardiovascular complications, such as retinopathy, nephropathy, and neuropathy in T2DM individuals (DECODE Study Group 2001; Milicevic et al. 2008). To reduce the progression of these complications, an optimal postprandial glycemic control should be considered in the treatment strategy (Del Prato 2002). Several reviews indicate that nutrition and lifestyle interventions can be effective in delaying the onset of the disease (Psaltopoulou et al. 2010; Thomas et al. 2010; Walker et al. 2010). Meta-analyses of randomized controlled trials reported that a low glycemic index (GI) diet has a clinically significant effect on glycemic control (Brand-Miller et al. 2003; Thomas and Elliott 2010). The GI, originally described by Jenkins et al. (1981), measures the extent to which

© The Author 2016
M. Ang, *Metabolic Response of Slowly Absorbed Carbohydrates in Type 2 Diabetes Mellitus*, SpringerBriefs in Systems Biology, DOI 10.1007/978-3-319-27898-8_1

carbohydrates affect blood glucose. Carbohydrates with a low GI are usually slowly absorbed, producing delayed gradual rises in blood glucose and insulin levels. However, the most recent nutritional guideline emphasizes quantity rather than quality of carbohydrates for the dietary management of T2DM (Evert et al. 2014).

Isomaltulose (ISO), an isomer of sucrose (SUC), is digested slower than other sugars such as SUC or maltose. Due to its GI value of 32 (Atkinson et al. 2008), it is classified as a low GI carbohydrate. A review of biological and toxicological studies reported that plasma glucose and insulin levels rise gradually after oral administration of ISO compared to SUC in healthy humans (Lina et al. 2002). Attenuations of postprandial plasma glucose and insulin levels following ISO versus SUC consumption have been demonstrated in healthy (Macdonald and Daniel 1983; Kawai et al. 1985, 1989; Liao et al. 2001; van Can et al. 2009; Holub et al. 2010; Maeda et al. 2013), impaired glucose-tolerant (van Can et al. 2012), and T2DM subjects (Kawai et al. 1989; Liao et al. 2001). The effects of ISO intake on postprandial glucose homeostasis, however, have not been comprehensively studied in T2DM individuals.

Dietary macronutrients carbohydrates, proteins, and fats are usually consumed in a complex food matrix rather than in their pure form. As a consequence, glycemic response may be modified because the macronutrients interact with each other. Understanding nutrient interdependencies is therefore important. For instance, food proteins alone stimulate insulin secretion, whereas a combined uptake of proteins with carbohydrates notably triggers additive effect on insulin release (Gannon et al. 1988, 1992; van Loon et al. 2003; Manders et al. 2005, 2006). The effects of insulin stimulation on glucose homeostasis, however, have been controversial in T2DM subjects. Some studies have shown an improvement in glucose response when proteins and carbohydrates were administered together compared to the uptake of carbohydrates alone (Gannon et al. 1988; Manders et al. 2005, 2006). Other studies did not confirm this effect (Gannon et al. 1992; van Loon et al. 2003). It is therefore unclear whether a combined load of protein and slowly digested ISO could improve postprandial glucose response in these subjects.

Based on the above considerations, this work aimed to firstly assess postprandial glucose metabolism after a bolus ingestion of ISO compared with SUC using a combined double tracer technique of euglycemic-hyperinsulinemic clamp and oral ISO or SUC load in T2DM subjects. This method enabled simultaneous quantification of systemic glucose appearance and disappearance rates accurately as well as differentiation between oral and endogenous glucose release into the blood circulation. Secondly, this work investigated whether a combined load of protein and slowly digested ISO could stimulate insulin secretion, thereby improving postprandial glucose response in T2DM patients. An isonitrogenous mixture of whey and soy proteins, as well as casein, was used as protein sources.

1.2 Contributions

The major contributions of the work in this book are:

- Conducting a systematic review on the effects of ISO consumption on post-prandial glucose metabolism in humans, including healthy probands, impaired glucose-tolerant subjects, and both type 1 diabetes mellitus (T1DM) and T2DM patients. This review provides a comprehensive up-to-date analysis of the evidence of attenuated plasma glucose and insulin levels after ISO ingestion compared with other carbohydrates such as SUC or dextrose.
- Simulating an appropriate glucose infusion (GINF) rate using pilot data of a hyperinsulinemic-euglycemic clamp in combination with oral ISO or SUC load in T2DM subjects. Under this experimental condition, blood glucose levels were aimed to remain in the near-normal range ($\sim$5 mmol/L) by infusing and adjusting GINF at a variable rate. The time-varying GINF rate was simulated by applying an algorithm developed by Furler et al. (1986).
- Analyzing and calculating postprandial glucose turnover rates extensively using stable isotope data. Under non-steady-state condition, postprandial glucose kinetics can be calculated by using two widely used approaches: Steele's one-compartment model (1-CM) (Steele 1959) and Mari's two-compartment model (2-CM) (Mari 1992). Considering the experimental condition applied in the hyperinsulinemic-euglycemic clamp, both models were slightly modified by combining them with the procedure described by Finegood et al. (1987). The results of both approaches were subsequently compared.
- Performing and assessing the feasibility of non-invasive methods for calculation of first-pass splanchnic glucose uptake (SGU). This parameter can be determined by using two available approaches: a tracer method or a method based on the procedure of Ludvik et al. (1997). The findings indicate that the first approach is accurate in estimating SGU, because SGU is calculated directly from ingested glucose tracer, leading to reproducible data. Conversely, the latter method is limited in its use because SGU is estimated indirectly, which depends exclusively on GINF data. Based on these considerations, SGU was calculated using labeled glucose tracer after administration of ISO or SUC.
- Establishing and implementing a method for estimation of tissue insulin sensitivity in the postprandial state. This was done by evaluating several modeling methods for determination of postprandial insulin resistance. The oral glucose minimal model is one of the approach that has been proposed to accurately measure overall insulin effects on glucose disposal and glucose production after glucose or meal ingestion in healthy subjects. This approach, which was originally developed by Bergman et al. (1979) and further extended by Caumo et al. (2000) and Dalla Man et al. (2002), has been validated by clamp experiment, tracer method, and intravenous glucose tolerance test (IVGTT) (Dalla Man et al. 2004, 2005; Cobelli et al. 2007). An insulin sensitivity index (S_I) was

derived from the general formulation of oral glucose minimal model, which was implemented in the estimation of insulin sensitivity following ISO or SUC administration in T2DM subjects.

1.3 Outline

This book begins with a theoretical background in Chap. 2 on the topic of diabetes mellitus, physiology of normal glucose homeostasis, and impairment of glucose homeostasis in T2DM in the postabsorptive and postprandial conditions. A state of the art reviews literature on the effects of ISO consumption on postprandial glucose metabolism in healthy, impaired glucose-tolerant, and diabetic subjects is provided. The last section of this chapter highlights implications from the literatures and develops work objectives. Chapter 3 explains the research method applied to address the work objectives proposed in Chap. 2. Research design, recruitment of subjects, and experimental procedure are outlined here. In addition, sample measurements, data analysis and calculations, and statistics are described extensively. Chapter 4 presents detailed results of the studies. Chapter 5 discusses the results and methodology used in the studies and draws conclusions based on the findings. Finally, the results are summarized in Chap. 6.

Chapter 2
Background and Objectives

2.1 Diabetes Mellitus

Diabetes mellitus describes a group of metabolic diseases, which is characterized by chronic hyperglycemia with metabolic disturbances of carbohydrate, lipid, and protein. It results from defects in insulin secretion, insulin action, or combination of both. Characteristic symptoms of hyperglycemia include polydipsia, polyuria, polyphagia, weight loss, and blurred vision. Severe hyperglycemia, if untreated, may lead to acute complications of developing potentially life-threatening ketoacidosis or hyperosmolar hyperglycemic state. Chronic hyperglycemia is associated with long-term complications of dysfunction and failure of different organs, such as retinopathy with potential blindness; nephropathy leading to renal failure; peripheral neuropathy with risk of foot ulcers, amputation, and neuropathic joints; and manifested autonomic neuropathy causing gastrointestinal symptom and sexual dysfunction as well as long-term risk of developing cardiovascular, peripheral vascular, and cerebrovascular diseases (WHO 1999; Genuth et al. 2003).

2.1.1 Diagnostic Criteria

Diabetes mellitus is diagnosed when measurements of fasting plasma glucose levels equal or exceed 7 mmol/L (126 mg/dL). If the estimated values lie in the uncertain range, i.e., between the cut-off values, the measurement should be repeated or alternatively an oral glucose tolerance test (OGTT) needs to be performed. Using OGTT, it is sufficient to measure plasma glucose level while fasting and at 2 h after a 75 g oral glucose load. The diagnosis is established when the 2-h plasma glucose level equals or exceeds 11.1 mmol/L (200 mg/dL). Both criteria have been suggested by WHO and the American Diabetes Association (ADA) over a decade ago (WHO 2006; ADA 2010). In 2009, an International Expert Committee

© The Author 2016
M. Ang, *Metabolic Response of Slowly Absorbed Carbohydrates in Type 2 Diabetes Mellitus*, SpringerBriefs in Systems Biology, DOI 10.1007/978-3-319-27898-8_2

consisting of representatives of ADA, European Association for the Study of Diabetes (EASD), and International Diabetes Federation (IDF) recommends a further method through measurement of glycated hemoglobin (HbA_{1c}) as a marker of chronic glycemia reflecting average blood glucose levels over a 2- to 3-month period of time (International Expert Committee 2009). This recommendation has been applied by ADA and WHO since 2010/2011. An HbA_{1c} threshold value of ≥ 48 mmol/mol (6.5%) confirms the diagnosis of diabetes mellitus (ADA 2010; WHO 2011b). In the absence of unequivocal hyperglycemia, one of the three methods should be re-tested until the diagnostic situation becomes clear. Additionally, patients with classical symptoms of hyperglycemia or hyperglycemic crisis can be diagnosed with diabetes mellitus when a random or casual plasma glucose level equals or exceeds 11.1 mmol/L (200 mg/dL) (ADA 2010).

People with impaired fasting glucose (IFG) and impaired glucose tolerance (IGT) are at increased risk to develop diabetes mellitus. The states of IFG and IGT are defined when plasma glucose levels lie in the intermediate range between normal and diabetic values. According to WHO, people diagnosed with IFG have fasting plasma glucose levels between 6.1 and 6.9 mmol/L (110–125 mg/dL) and those with IGT have 2-h plasma glucose values between 7.8 and 11.1 mmol/L (140–200 mg/dL) after a 75 g OGTT (WHO 2006). The ADA applies a slightly different range for IFG with fasting plasma glucose levels of 5.6–6.9 mmol/L (100–125 mg/dL) (ADA 2010). Table 2.1 summarizes the diagnostic criteria for diabetes and intermediate hyperglycemia.

Table 2.1 Criteria for diagnosis of diabetes mellitus and intermediate hyperglycemia

	Fasting glucose mmol/L (mg/dL)	2-h glucose mmol/L (mg/dL)	HbA_{1c} mmol/mol (%)
IFG/IGT (WHO)	6.1–6.9 (110–125)	7.8–11.1 (140–200)	Undefined
IFG/IGT (ADA)	5.6–6.9 (100–125)	7.8–11.0 (140–199)	39–46 (5.7–6.4)
Diabetes mellitus (WHO)	≥ 7.0 (126)	≥ 11.1 (200)	≥ 48 (6.5)
Diabetes mellitus (ADA)	≥ 7.0 (126)	≥ 11.1 (200)	≥ 48 (6.5)

Values represent venous plasma glucose. IFG and IGT are the abnormalities of glucose regulation in the fasting and postprandial states, respectively. In the absence of unequivocal hyperglycemia, the above criteria for diagnosis of diabetes should be confirmed by repeated measurements. Recommended criteria according to WHO (2006), ADA (2010), and WHO (2011b)

2.1.2   Classification

Diabetes mellitus is categorized into four types, namely T1DM, T2DM, gestational diabetes mellitus (GDM), and other specific types of diabetes. Among those types, T1DM and T2DM are the two diabetic cases that appear frequently. T1DM begins mostly in the early childhood; people affected by T1DM have usually normal weight. T2DM, on the other hand, is more common in older adults who are obese, but nowadays it has become increasingly prevalent in obese children and adolescent (Reinehr 2013).

T1DM, formerly known as insulin-dependent diabetes or juvenile-onset diabetes, accounts for 5–10 % of the cases and is characterized by loss of insulin-producing beta-cells of the islets of Langerhans in the pancreas, leading to absolute insulin deficiency. This is caused by an autoimmune destruction of the pancreatic β-cells, which can be identified by using markers of immune-mediated destruction of β-cells, such as autoantibodies to islet cells, insulin, glutamic acid decarboxylase, and tyrosine phosphatases islet antigen 2 and 2β in combination with genotyping of human leukocyte antigen DR and DQ gene loci. The autoimmune β-cells destruction is related to genetic predisposition and may also be triggered by environmental factors that are however still insufficiently defined. Due to lack of insulin, hyperglycemia becomes manifest with a typical symptom of ketoacidosis. In advanced stage of T1DM, endogenous insulin secretion remains little or none as identifiable by the low or undetectable plasma C-peptide levels (WHO 1999; ADA 2010). As a result, exogenous insulin administration becomes necessary for controlling the manifested hyperglycemia. Insulin therapy often includes the use of insulin analogues and mechanical technologies such as insulin pumps and continuous glucose monitors (Atkinson et al. 2014). Hypoglycemia may occur with intensified insulin treatment, but mostly is due to defective glucose counterregulation (Ang et al. 2014).

T2DM, formerly known as non-insulin-dependent diabetes or adult-onset diabetes, is the most common form of diabetes, accounting for 90–95 % of the cases. Individuals with T2DM are usually insulin resistant and have a relative insulin deficiency in opposite to T1DM. By definition, autoimmune destruction of β-cells does not occur, and none of the above other classified types of diabetes mellitus causes this form of diabetes (WHO 1999; ADA 2010). Its etiology is complex and involves multiple processes, including genetic, social, behavioral, and environmental factors (Ripsin et al. 2009; Reinehr 2013). The latter one is thought to be the predominant cause with obesity representing the hallmark of the disease. Most individuals with T2DM are overweight or obese, defined as having a body mass index (BMI) greater than 25, whereas those who are non-obese may have a marked increase of abdominal body fat. Obesity itself may also cause insulin resistance. This type of diabetes remains often undiagnosed even after several years of manifested hyperglycemia, because the existing hyperglycemia is not sufficiently severe to cause remarkable symptoms. Ketoacidosis occurs seldom in T2DM. Plasma insulin and C-peptide levels are elevated in contrast to T1DM. Most of the evidence-based guidelines for T2DM management focus primarily on lifestyle modification, normalization of blood glucose level, and reducing the risk factors for micro- and macrovascular complications. Lifestyle intervention such as weight reduction, increased physical activity, and reduced calorie as well as fat intake may help improve insulin resistance. If compliances to diet and exercise are low or the glycemic goal (HbA$_{1c}$<7 %) is not achieved, pharmacological therapy is required. Treatment with insulin is introduced when glycemic control is no longer possible with oral agents or when contraindication to oral medications exists (WHO 1999; ADA 2010; Ripsin et al. 2009; Reinehr 2013).

The third type, i.e., GDM, is hyperglycemia that is first recognized during pregnancy. Other types of diabetes mellitus can be caused by monogenetic defects in β-cell function, such as impaired insulin secretion due to mutations on different chromosomes (e.g., defects in hepatic transcription factor or glucokinase gene), commonly referred to as maturity-onset diabetes of the young (MODY) with minimal or no defects in insulin action; mutations in insulin receptor gene that cause defects in insulin action; diseases of exocrine pancreas (e.g., pancreatitis, pancreatectomy, pancreatic carcinoma); excess of hormones, i.e., growth hormone, cortisol, glucagon, epinephrine which antagonize insulin action (e.g., acromegaly, Cushing's syndrome, glucagonoma, pheochromocytoma, respectively); drugs that impair insulin secretion and insulin action; as well as infections by certain viruses associated with destruction of β-cells (WHO 1999; ADA 2010).

2.2 Glucose Homeostasis

Plasma glucose levels fluctuate throughout the day as a result of increasing or decreasing supplies (e.g., during fasting, eating, or exercise) but are restored within a narrow range at approximately 5 mmol/L (90 mg/dL). The maintenance process of plasma glucose at constant concentration (normoglycemia) is termed glucose homeostasis, resulting from coordination of factors that regulate the rate of glucose entering the circulation (rate of appearance/release, R_a) and the rate of glucose leaving the circulation (rate of disappearance/disposal, R_d). By definition, a normal fasting plasma glucose level should be lower than 6.1 mmol/L (110 mg/dL), and a normal 2-h plasma glucose level should not exceed 7.8 mmol/L (140 mg/dL) (WHO 2006).

2.2.1 Role

Glucose is an essential source of cellular energy. While most tissues can utilize free fatty acids (FFA) as a metabolic fuel in addition to glucose, the brain is critically dependent on the constant supply of glucose from plasma. This occurs due to several limitations; FFA do not cross the blood-brain barrier, brain cannot synthesize or store glucose, and ketone bodies as alternative substrates are normally present at low levels in plasma. During starvation period, ketone bodies may become important substitutes for glucose because their circulating concentrations are increased (Gerich 2000). Conditions of hypoglycemia, as plasma glucose concentrations fall below the physiological range (<2.8 mmol/L, 50 mg/dL) or more severe forms, cause impairment in brain function, brain damage, and even death (Siesjö 1988; Cryer 2008). Conversely, elevated plasma glucose levels are associated with increased risks for cardiovascular disease and mortality (DECODE Study Group 2003; Levitan et al. 2004). Therefore, glucose homeostasis is fundamental for living organisms in order to prevent from pathological consequences resulting from hypoglycemia or hyperglycemia.

2.2.2 Regulatory Factors and Actions

Regulation of glucose homeostasis involves control of endogenous glucose production (EGP) and whole-body glucose utilization by various tissues, including liver, kidney, skeletal muscle, adipose tissue, gastrointestinal tract, pancreas, and brain. Among them, the liver plays a major role. It releases glucose into the circulation in the fasting state, takes up some of the ingested glucose derived from meal, and stores them as glycogen. The liver is also able to breakdown glycogen into glucose (glycogenolysis) and synthesize glucose from the substrates lactate, amino acids, and glycerol (gluconeogenesis). Thus, the liver serves to keep a constant plasma glucose level by modulating glucose uptake as well as glucose production via glycogenolysis and gluconeogenesis (Wahren and Ekberg 2007).

Besides the liver, the kidney is the only organ capable of producing glucose. Both liver and kidney contain the required enzyme glucose-6-phosphatase for releasing glucose into the blood. Due to lack of this enzyme, many other organs are limited in their function to release glucose. Unlike the liver, the kidney contains only little amount of glycogen, and therefore it synthesizes and releases glucose almost exclusively through gluconeogenesis. Although its contribution to overall systemic glucose is small, the proportion of glucose production due to gluconeogenesis is roughly comparable to that of the liver (Gerich 2000). Other renal regulatory mechanisms include uptake of glucose from the circulation to meet its energy needs, reabsorption of glucose at the proximal tubule, and elimination of excess glucose in the urine (Triplitt 2012b).

The majority of peripheral glucose disposal occurs in muscle cells, with a small amount taken up by the adipose tissues. After entry into cells, glucose may be immediately oxidized for energy (oxidative glycolysis), converted to lactate or alanine (non-oxidative glycolysis), or stored as glycogen or lipid. Skeletal muscle cannot release free glucose into circulation; however, it has the ability to produce lactate and amino acids from its glycogen storage and protein pool. Lactate, alanine, and glutamine are the major precursors to be incorporated into glucose by the liver and kidney (Gerich 2000; Meyer et al. 2005). Although adipocytes contribute only to a small proportion of the total body glucose disposal, it plays an important role in the maintenance of glucose homeostasis by regulating FFA release from stored triglycerides and influencing insulin sensitivity in muscle and liver (Triplitt 2012a).

Multiple glucoregulatory hormones are involved in the regulation of glucose homeostasis. Insulin and glucagon are the most relevant ones. Both hormones are produced in the pancreas by islet of Langerhans; β-cells secrete insulin while α-cells release glucagon. Insulin acts to reduce plasma glucose levels through several mechanisms. It increases glucose disposal into insulin-sensitive tissues (muscle and adipose cells), which facilitates glycogenesis (glycogen synthesis) and lipogenesis (fat formation) (Triplitt 2012a). Further, it inhibits glucose release from liver and kidney directly via enzyme activation or deactivation and indirectly via suppression of gluconeogenic substrate availability and glucagon secretion (Gerich 2000). Insulin reduces FFA release by inhibiting lipolysis (fat breakdown) and promoting triglyceride storage and simultaneously increases FFA clearance from the circulation. In addition, it stimulates amino acids uptake and protein synthesis in muscle cells, thereby reducing the circulating substrates for gluconeogenesis (Meyer et al. 1998a; Aronoff et al. 2004). Glucagon also plays a central role in glucose homeostasis by acting exclusively on the liver to enhance hepatic glycogenolysis, which decelerates after several hours and is followed by increased gluconeogenesis (Magnusson et al. 1995; Gerich 2000). Glucagon secretion is stimulated in response to hypoglycemia, leading to increased hepatic glucose production, and is inhibited by hyperinsulinemia (Aronoff et al. 2004; Triplitt 2012a). Thus, glucagon exhibits the opposite effects of insulin through elevation of plasma glucose levels.

Incretin hormones are also involved in the regulation of glucose homeostasis. Glucose-dependent insulinotropic polypeptide (GIP) and glucagon-like peptide-1 (GLP-1) are the two incretin hormones that are secreted from the small intestine in response to meal ingestion. Both hormones regulate insulin secretion within the nutrients in the food, also called the incretin effect. The concept of incretin was used to explain a greater insulin response induced by oral glucose load compared to the same amount of intravenous glucose administration. GIP is secreted from enteroendocrine K-cells primarily in the upper part of the small intestine to facilitate glucose disposal following meals containing carbohydrate or fat, enhance incorporation of FFA into triglyceride, and modulate FFA synthesis. On the other hand, GLP-1 is released from enteroendocrine L-cells located in the small bowel and lower gut, which acts in addition to its insulinotropic effect to inhibit gastric emptying, suppress glucagon secretion, and reduce EGP; all of which help to lower blood glucose levels (Kim and Egan 2008).

A group of hormones acts to regulate glucose homeostasis once plasma glucose levels fall below the normal value. Due to the brain dependence on plasma glucose, these hormones are activated to prevent hypoglycemia; they are called counterregulatory hormones, which include glucagon, catecholamines (epinephrine and norepinephrine), cortisol, and growth hormone. Collectively, these hormones rise plasma glucose levels by increasing EGP and decreasing glucose uptake by insulin-sensitive tissues (Ang et al. 2014).

2.2.3 Regulation of Glucose Homeostasis

2.2.3.1 Postabsorptive State

In the postabsorptive state (10–16 h overnight fast), plasma glucose levels are relatively stable ($\sim$5 mmol/L) since rates of glucose release into the circulation approximate rates of glucose removal from the circulation ($\sim$10 μmol/kg/min). During this period, $\sim$80–85 % of glucose production is derived from the liver, and the remaining $\sim$15–20 % originates from the kidney (Gerich 2000; DeFronzo 2004). Glycogenolysis and gluconeogenesis contribute equally to the overall glucose release; approximately half of the amount of glucose produced is the result of hepatic glycogenolysis, and another half is due to gluconeogenesis with approximately similar proportion from both liver and kidney. When the fasting period is prolonged and hepatic glycogen stores become depleted, the contributions of both hepatic and renal gluconeogenesis increase even more (Gerich 2000; Wahren and Ekberg 2007).

Glucoregulatory hormones control hepatic and renal glucose production differently. Glucagon exerts its effect solely in the liver. Thus, in the postabsorptive state, more than half of the glucose derived from hepatic glycogenolysis and gluconeogenesis is dependent on the maintenance of normal basal glucagon levels. Epinephrine increases glucose release predominantly via renal gluconeogenesis and to a lesser extent through increased concentrations of gluconeogenic substrates (Gerich 2000). Insulin, on the other hand, suppresses glucose release by both liver and kidney. However, this action is minimal due to low insulin secretion in the fasting state (Aronoff et al. 2004).

The major source of body's energy requirements in the postabsorptive state is matched by oxidation of FFA. Several tissues, such as brain and red blood cells, with an obligatory need for glucose, yet continue to consume it. To supply these organs with glucose, splanchnic tissues (liver and gut), mainly liver, therefore switch from glucose uptake to glucose production (Wahren and Ekberg 2007). Consequently, most of the circulating glucose in the fasting state is used by the brain ($\sim$45–60 %), and only a small proportion is taken up by the splanchnic tissues ($\sim$3–6 %). Postabsorptive glucose disposal also takes place in other organs, such as in muscle cells ($\sim$15–20 %), kidney ($\sim$10–15 %), blood cells ($\sim$5–10 %), and adipose tissues ($\sim$2–4 %) (Gerich 2000). Glucose disposal rates into muscle and adipose tissues depend on basal insulin levels in the postabsorptive state (Consoli et al. 1992; DeFronzo 2004), whereas glucose uptake by brain, renal medulla, splanchnic tissues, and blood cells occurs largely independent on insulin. Rather, it depends on tissue demands, mass action effect of plasma glucose concentration per se, and the number and characteristics of glucose transporters (GLUTs) in specific tissues. After its uptake into cells, glucose is not stored as glycogen but completely oxidized to CO_2 or undergoes non-oxidative glycolysis and is released back into the circulation as lactate, alanine, and glutamine for further resynthesis into glucose (Gerich 2000). Regulation of postabsorptive glucose homeostasis is summarized in Fig. 2.1.

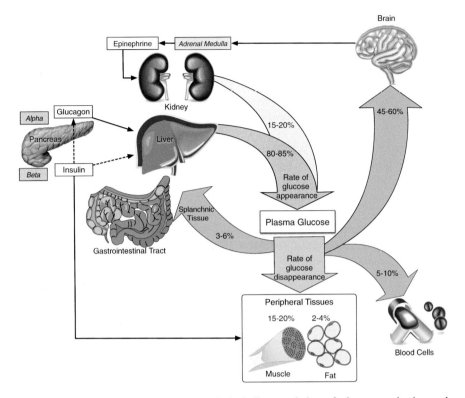

Fig. 2.1 Postabsorptive glucose homeostasis including regulation of glucose production and glucose utilization by several tissues as well as glucoregulatory hormones. Glucose production is derived mainly from the liver under the control of glucagon, with a little contribution from the kidney. Glucose uptake in muscle and adipose tissues is dependent on basal insulin levels, whereas its disposal into other organs occurs largely independent of insulin. Insulin action on glucose production is decreased due to low insulin secretion in the fasting state. *Solid arrows* represent enhanced effect, and *dashed arrows* represent diminished effect (Percentage values adapted from Gerich 2000 and DeFronzo 2004)

2.2.3.2 Postprandial State

The magnitude of circulating glucose excursions following meal ingestion can be affected by several factors that enter into and exit from the systemic circulation. These include food digestion within the lumen of the small intestine, glucose absorption into the portal vein, glucose extraction by the splanchnic tissues, endogenous glucose release by the liver and kidney, and posthepatic glucose uptake with subsequent storage, oxidation, or non-oxidative glycolysis (Gerich 2000). Following digestion and absorption of meal-derived glucose from the small intestine, glucose is transported via the portal vein to the liver. During this time, the liver switches from glucose production to glucose uptake and storage (Ludvik et al. 1997; Wahren and Ekberg 2007). A portion of the ingested glucose is initially metabolized by the splanchnic tissues, and the remainder reaches the systemic circulation. Thus, under postprandial condition, the amount of glucose released into the blood circulation is determined by the rates of glucose absorption, SGU, and EGP. Rates of systemic glucose appearance in plasma, therefore, represent the sum of orally ingested glucose escaping first-pass splanchnic extraction and the endogenous glucose released by the liver and kidney (Gerich 2000; Basu et al. 2001). Ingested glucose appears in the circulation as early as 15 min, reaches a peak, and decreases gradually thereafter. At the same time, EGP is suppressed (Gerich 2000; Aronoff et al. 2004; Wahren and Ekberg 2007).

Approximately 30 % of ingested glucose is initially extracted by the splanchnic tissues (mostly liver) and may be immediately stored as glycogen (Gerich 2000; Wahren and Ekberg 2007). The remaining glucose, which enters the systemic circulation, is mainly taken up by muscle cells ($\sim$35–40 %), again the liver ($\sim$20 %), and other tissues, such as brain ($\sim$20 %), kidney ($\sim$10 %), adipocytes and other cells ($\sim$7–15 %) (Gerich 2000). Glucose taken up by the tissues is initially oxidized and later is stored as glycogen or may undergo non-oxidative glycolysis, leading to formation of 3-carbon compounds, such as pyruvate, lactate, and alanine. These compounds may be synthesized into glucose via gluconeogenesis, subsequently may either be stored in glycogen or be released into plasma as glucose (Woerle et al. 2003).

Plasma glucose levels are determined by the balance between the rates of systemic glucose appearance and the rates of systemic glucose disappearance. During the first 2 h following meal ingestion, plasma glucose levels increase but rarely beyond 9 mmol/L, because rates of glucose appearance exceed rates of glucose disappearance in the systemic circulation. Glucose disappearance rates normally follow a similar pattern to the rates of glucose appearance but are shifted in time (Gerich 2000). Elevated plasma glucose levels necessitate insulin secretion to increase glucose transport into muscle and adipose tissues. Endogenous glucose release is concomitantly suppressed via direct action of insulin by binding to its receptors that activates insulin-signaling pathway, and via paracrine effect or direct communication within the pancreas between α- and β-cells, resulting in inhibition of glucagon secretion (Aronoff et al. 2004). The coordinated reciprocal release of insulin and glucagon as well as early insulin rise predominantly determine the suppression of glucose release and the initial splanchnic glucose extraction (Gerich 2000). Plasma FFA and glycerol concentrations decrease due to the inhibition of lipolysis, while plasma lactate concentrations increase as a result of increased glycolysis (Woerle et al. 2003). In addition to the insulin release, gut incretin GLP-1 and GIP hormones are secreted to facilitate glucose-dependent insulin secretion. Figure 2.2 summarizes regulation of postprandial glucose homeostasis in healthy persons.

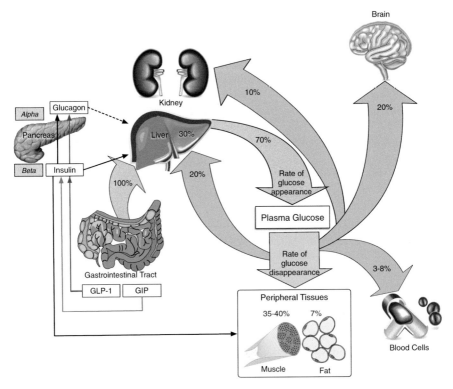

Fig. 2.2 Postprandial glucose homeostasis including regulation of glucose release and glucose utilization by several tissues as well as glucoregulatory hormones. After meal ingestion, glucose is absorbed from the small intestine and transported via the portal vein to the liver. A portion of glucose is initially taken up by the liver, and the remainder is released into the circulation. Insulin secretion is increased, leading to inhibition of glucose production and enhancement of glucose disposal. Additionally, gut incretin GLP-1 and GIP hormones are stimulated to facilitate glucose-dependent insulin secretion. *Solid arrows* represent enhanced effect, and *dashed arrows* represent diminished effect (Percentage values adapted from Gerich 2000)

2.2.4 *Abnormal Glucose Homeostasis in Type 2 Diabetes Mellitus*

Glucose homeostasis is impaired in T2DM, which manifests in fasting and post-prandial hyperglycemia. Multiple factors contribute to the development of abnormal glucose homeostasis in T2DM, with the central defects being defective insulin secretion (insulin deficiency) and diminished tissue insulin sensitivity (insulin resistance) (WHO 1999; ADA 2010).

2.2.4.1 Pathogenesis

Insulin Deficiency Defective insulin secretion is a characteristic feature in T2DM individuals. Early in the course of T2DM, insulin resistance is manifested, but glucose tolerance remains normal due to compensatory increase in insulin secretion. Elevated insulin release is marked by increased basal C-peptide level in T2DM, while fasting plasma insulin level is normal or increased. The relationship between fasting plasma glucose and insulin concentrations represents an inverted U-shape curve. When fasting plasma glucose levels increase to 7.8 mmol/L (140 mg/dL), fasting plasma insulin levels rise substantially above the normal values. This can be viewed as an adaptive response of pancreas to maintain glucose homeostasis. With time, the high insulin secretion cannot be maintained, β-cell starts to fail. A further overshoot in fasting glucose levels will cause deterioration in β-cell function, resulting in diminished fasting insulin concentrations. At this stage of established impaired insulin secretion, hyperglycemia occurs; basal hepatic glucose production begins to rise (DeFronzo 2004).

Also, there exists an inverted U-shaped relationship between insulin response after a post-glucose challenge and fasting glucose level in T2DM. Postprandial plasma insulin response begins to decline as fasting plasma glucose level increases to ~6.7 mmol/L (120 mg/dL). Typically, a T2DM individual with a fasting glucose level of 8.3–8.9 mmol/L (150–160 mg/dL) secretes as much insulin as that in healthy subjects. However, in the presence of hyperglycemia and insulin resistance, this response is markedly reduced. As a result, T2DM exhibits elevated fasting but decreased postprandial insulin and C-peptide secretion (DeFronzo 2004).

Early in the development of diabetes, individuals with IGT have already lost 60–70 % of their β-cell function. A defect in the first-phase insulin release is observed in T2DM subjects after an IVGTT or an OGTT. Loss of the early insulin secretion during the IVGTT (0–10 min) and OGTT (0–30 min) becomes particularly evident as fasting glucose levels exceeding 6.1–6.7 mmol/L (110–120 mg/dL). The defect is most obvious during the OGTT if the incremental plasma insulin level at 30 min is expressed relative to the incremental plasma glucose level at 30 min ($\Delta I_{30}/\Delta G_{30}$). This disorder has important pathogenic consequences, because early insulin release primes insulin target tissues to maintain normal glucose homeostasis (DeFronzo 2004).

Insulin Resistance Insulin resistance contributes substantially to the development of T2DM. Several studies have consistently demonstrated a diminished insulin action in T2DM individuals using multiple techniques, such as combined oral glucose and intravenous insulin tolerance test, insulin suppression test, arterial insulin infusion into forearm and leg muscles in combination with (radio)isotope tracers, frequently sampled IVGTT, and minimal model technique (DeFronzo 2004). This impairment is characterized by insulin's inability to reduce plasma glucose level through suppression of glucose production and enhancement of tissue glucose uptake. By means of euglycemic hyperinsulinemic clamp technique, whole-body glucose disposal in lean and overweight diabetic subjects is reduced ~30–50 %

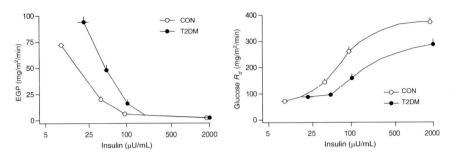

Fig. 2.3 EGP and glucose R_d in response to changes in insulin in 14 T2DM patients and 14 normal healthy volunteers. Data are mean $\pm$ SEM. Start point represents baseline levels. The other 3 points on line are results of sequent insulin infusion at rates of 0.4, 1.0, and 10 mU/kg/min. The first rate was infused for 3 h, the other two rates for 2 h. *CON* healthy control subjects (Graphs adapted from Campbell et al. 1988)

compared to that of healthy volunteers. This defect is observed at all plasma insulin concentrations (physiologic and pharmacologic ranges) (Campbell et al. 1988; DeFronzo 2004). EGP is less suppressed at physiological insulin concentrations ($\sim$40–100 μU/mL) in these subjects compared to normal healthy probands, but can be completely inhibited by a level of $\sim$2000 μU/mL (Fig. 2.3) (Campbell et al. 1988). In diabetic subjects with severe fasting hyperglycemia, even maximal plasma insulin concentrations fail to restore normal glucose tolerance (DeFronzo 2004).

The relative contribution of insulin deficiency and insulin resistance to the manifestation of hyperglycemia may differ in T2DM because of the heterogeneity of the disease. However, in most cases, insulin resistance precedes the development from normal glucose tolerance to IGT and overt diabetes in all ethnic populations. Numerous studies have consistently shown that progression from normal glucose tolerance to IGT starts with severe insulin resistance, which is then largely counterbalanced by increased insulin secretion. In the following sequence, insulin secretion becomes gradually or markedly reduced, whereas insulin resistance is slightly diminished or remain undeteriorated. During the transition from normal glucose tolerance to diabetes, insulin sensitivity deteriorates $\sim$40 %, whereas insulin secretion decreases by 3- to 5-fold. Consequently, glucose homeostasis is impaired, resulting in fasting and postprandial hyperglycemia. Chronic hyperglycemia in T2DM may in turn exacerbate both insulin deficiency and insulin resistance, which are further aggravated by chronic elevation in plasma FFA levels (DeFronzo 2004; Triplitt 2012a).

2.2.4.2 Metabolic Alterations

Postabsorptive State The majority of tissue glucose disposal occurs independent of insulin in the postabsorptive state. Approximately 25 % of glucose taken up by peripheral muscle and adipose tissues is mediated by insulin, consequently only a relatively small proportion of overall postabsorptive glucose disposal is affected by the conditions of insulin resistance and insulin deficiency in T2DM. Because rates of glucose production are increased and most of glucose released in the fasting condition is produced by the liver, increased hepatic glucose output is considered to contribute substantially to postabsorptive hyperglycemia in T2DM. A direct relationship between basal hepatic glucose production and fasting glucose concentrations has been observed ($r = 0.83$–0.85, $P < 0.001$). Hepatic glucose production is sensitive to insulin action. Since fasting insulin levels are elevated in T2DM and hyperinsulinemia is a potent inhibitor of hepatic glucose release, the presence of hepatic insulin resistance can explain the excessive basal glucose output. In addition, the liver is resistant to the mass action of hyperglycemia because hyperglycemia per se exerts a powerful inhibitory action on glucose release (Consoli 1992; DeFronzo 2004; Wahren and Ekberg 2007).

The augmented basal hepatic glucose flux in T2DM can be derived from enhanced glycogenolysis, gluconeogenesis, or both. Studies using nuclear magnetic resonance spectroscopy indicate that both hepatic glycogen storage and glycogenolysis rate are reduced in T2DM patients with poor metabolic control but remain normal in those with good metabolic control. On the contrary, hepatic gluconeogenesis is reported to be elevated in these patients, and increasing evidence suggests that increased gluconeogenesis rather than glycogenolysis is the predominant process responsible for the excessive hepatic glucose release in the postabsorptive state in T2DM. Several factors contribute to accelerate the process, including increased circulating gluconeogenic substrates (lactate, pyruvate, glycerol, alanine, and other amino acids), enhanced hepatic substrate extraction, and elevated intrahepatic substrate conversion into glucose. Besides insulin resistance and insulin deficiency, other mechanisms have been shown to also cause the increased efficiency of hepatic gluconeogenesis, such as hyperglucagonemia and excessive hepatic FFA oxidation consequence to the increased lipolysis (Consoli 1992; Gerich 1993; DeFronzo 2004; Wahren and Ekberg 2007).

Figure 2.4 depicts the pathophysiological mechanisms of fasting hyperglycemia in T2DM. The combination of hepatic insulin resistance, impaired insulin secretion, hyperglucagonemia, and increased FFA oxidation act together to promote hepatic gluconeogenesis, causing a rise in hepatic glucose release. As a result, plasma glucose levels increase to a point where there is a compensatory increase in insulin secretion. Hyperinsulinemia and the mass action of hyperglycemia compensate for insulin resistance in the peripheral tissues, and glucose disposal increases to balance the elevated circulating glucose release. Consequently, glucose homeostasis reaches a new steady state at a greater fasting plasma glucose level. The increased glucose uptake by peripheral tissues in turn provokes glycolysis, which elevates gluconeogenic substrate supply for hepatic gluconeogenesis and causes an increased rate of hepatic glucose production to be maintained (Consoli 1992).

The elevated systemic postabsorptive rates of glucose disappearance have been consistently shown in T2DM (Campbell et al. 1988; Consoli 1992; Gerich 1993; DeFronzo 2004). Although glucose uptake by peripheral insulin-sensitive tissues contributes only to a small proportion of the overall glucose disappearance in the

Fig. 2.4 Mechanisms of postabsorptive hyperglycemia in T2DM (Schema adapted from Consoli 1992)

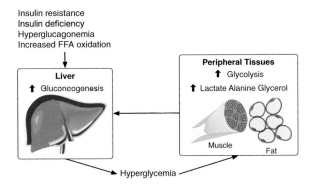

fasting state, reduced efficiency of glucose removal due to both insulin deficiency and insulin resistance may at least participate in determining the magnitude of the achieved steady-state hyperglycemia. Approximately 80 % of insulin-stimulated glucose disposal takes place in muscle cells, and postabsorptive muscle glucose uptake has been reported to be increased in absolute terms. However, glucose clearance rate, i.e., the efficiency rate of glucose uptake by muscle cells, is diminished (Gerich 1993; DeFronzo 2004).

Postprandial State Plasma glucose levels increase excessively in T2DM subjects following an oral glucose load or a meal ingestion. Mechanistically, elevated postprandial hyperglycemia occurs because rates of glucose appearance into the circulation markedly exceed rates of glucose disappearance. Since absolute rates of glucose removal and muscle glucose uptake are normal or increased in T2DM, enhanced glucose appearance in plasma contributes substantially to the abnormality in postprandial hyperglycemia in these subjects. This could result from failure to adequately suppress endogenous glucose output, diminished splanchnic glucose sequestration, or combination of both, because appearance of ingested glucose in the circulation is generally normal (Consoli 1992; Gerich 1993; Wahren and Ekberg 2007).

Several mechanisms may explain the impaired suppression of endogenous glucose release in T2DM. These include delayed and decreased insulin secretion relative to prevailing plasma glucose level, impaired suppression of glucagon secretion, and presence of hepatic insulin resistance. Similar to postabsorptive state, accelerated hepatic gluconeogenesis may represent an important mechanism for the increased glucose production following carbohydrate or meal ingestion. Increased availability of gluconeogenic substrates lactate and alanine secondary to defects in muscle glycogen storage may fuel increased postprandial gluconeogenesis (Consoli 1992; Gerich 1993). Incretin deficiency or resistance to incretins may also play a role. In T2DM subjects, GIP response to glucose ingestion is normal whereas GLP-1 response is reduced. In these subjects, however, insulinotropic effect of GIP is blunted while the effect of GLP-1 is preserved (DeFronzo 2004; Holst and Gromada 2004; Kim and Egan 2008). In addition to increased glucose release, there is evidence suggesting that a decrease in SGU following an oral glucose load may explain the elevated postprandial hyperglycemia in these subjects (Ludvik et al. 1997; Basu et al. 2001).

Peripheral glucose uptake and muscle glucose disposal of T2DM are apparently normal in postprandial state in absolute terms. This absolute value is however inappropriately low when considering the prevailing plasma glucose concentrations in T2DM. Studies in T2DM subjects have shown reduced overall peripheral glucose clearance as well as muscle glucose clearance after glucose ingestion. Although muscle glucose uptake in T2DM subjects is comparable to that of normal individuals following glucose ingestion, the metabolic fate of glucose in muscle cells is nevertheless abnormal. Glucose oxidation is low or inappropriately normal. Muscle glucose storage is reduced, causing an increase in muscle net release of lactate, pyruvate, and alanine. Thus, the abnormalities in glucose disposal can also contribute to excessive postprandial hyperglycemia in T2DM (Consoli 1992; Gerich 1993).

2.3 Glycemic Control

Increasing evidence suggests that diabetic individuals with larger blood glucose fluctuation within the day and from day-to-day basis may have a greater risk to develop long-term cardiovascular complications. To reduce the progression of T2DM, an optimal glycemic control strategy should therefore be considered in the treatment. This includes controlling fasting and postprandial hyperglycemia and reducing HbA_{1c} level as an index of chronic glycemia (Del Prato 2002). Maintaining glucose levels closely to normal range has been shown to reduce diabetes-related microvascular complications, including retinopathy, nephropathy, and neuropathy (Nathan et al. 2009).

2.3.1 Glycemic Goals of Therapy

The glycemic goal recommended by ADA and EASD is to achieve and maintain HbA_{1c} level of <7%, with an emphasis on fasting glucose. Target fasting and preprandial levels of plasma and capillary glucose should be in the range of 3.9–7.2 mmol/L (70–130 mg/dL) (Nathan et al. 2009). Controlling fasting hyperglycemia is necessary; however, it is usually insufficient to obtain optimal glycemic control. Epidemiological reports suggest that reducing postprandial plasma glucose excursions is as important, or perhaps more important for achieving HbA_{1c} goal (DECODE Study Group 2001; Milicevic et al. 2008). Recently, the IDF recommends a target postprandial glucose level of 9 mmol/L (160 mg/dL) that should be measured 1–2 h after a meal (IDF 2014).

2.3.2 Low Glycemic Index Diet for Glycemic Control

Several reviews reported that nutrition and lifestyle interventions can be effective in delaying the onset of diabetes (Psaltopoulou et al. 2010; Thomas et al. 2010; Walker et al. 2010). Meta-analyses of randomized controlled trials reported that a low GI diet has a clinically relevant effect on glycemic control (Brand-Miller et al. 2003; Thomas and Elliott 2010). The GI, originally described by Jenkins et al. (1981), measures the extent to which carbohydrates affect blood glucose. Following consumption of a 50 g carbohydrate from a test food, the area under postprandial plasma glucose curve is calculated and compared to that following of the same amount of carbohydrate from a standard food (glucose or white bread). The GI of the test food is expressed as percentage of the standard food. Starchy staples foods of traditional cultures have often lower GIs, such as pasta, whole-grain pumper-nickel breads, cracked wheat or barley, rice, dried peas, beans, and lentils (Jenkins et al. 2002).

The GI concept is an extension of the dietary fiber hypothesis, which suggests that fiber may delay absorption of nutrients within the small intestine. It is hypothesized that the metabolic effect of low GI foods relates to the rates at which glucose is absorbed from the small bowel. Low GI foods would leave the stomach and travel along the small intestine where they are digested and absorbed more slowly. On the other hand, high GI foods would be released from the stomach and be rapidly absorbed high up in the gut. A reduction in the rate of glucose absorption will cause gradual rise in blood glucose levels, whereas more rapidly absorbed glucose will result in an undershoot of blood glucose (Fig. 2.5). Slowly absorbed carbohydrates would have beneficial effect in the treatment of metabolic disease such as diabetes where gradual blood glucose rises are essential (Jenkins et al. 1987, 2002).

Fig. 2.5 Hypothetical effects of consumption of carbohydrates with a low or a high GI on gastrointestinal glucose absorption and postprandial blood glucose concentration (Figure adapted from Jenkins et al. 1987)

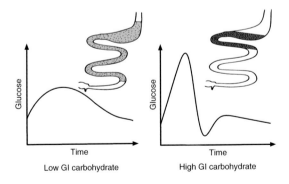

2.3.3 *Isomaltulose*

ISO is an example of carbohydrate with a low GI (GI $=$ 32) (Atkinson et al. 2008). ISO is a disaccharide occurring naturally as a minor component in honey and sugar cane extract (Siddiqui and Furgala 1967; Takazoe 1985). It is present in honey in a quantity of about 0.35 % (Low and Sporns 1988). ISO, also called palatinose, has the chemical name 6-O-α-D-glucopyranosyl-D-fructofuranose. ISO possesses the same chemical formula as SUC, i.e., $C_{12}H_{22}O_{11}$ but different structures, and therefore is a SUC isomer. As shown in Fig. 2.6, both disaccharides are formed by linkage of 2 monosaccharides (glucose and fructose). Both glucose and fructose molecules are connected by α-1,6 glycosidic bond in ISO, instead of α-1,2 in SUC. Commercially, ISO is produced from SUC by enzymatic rearrangement of the glycosidic linkage from α-1,2 to α-1,6 and followed by crystallization (Lina et al. 2002). The enzyme used in the conversion is usually obtained from a nonpathogenic microorganism Protaminobacter rubrum (Porter et al. 1991).

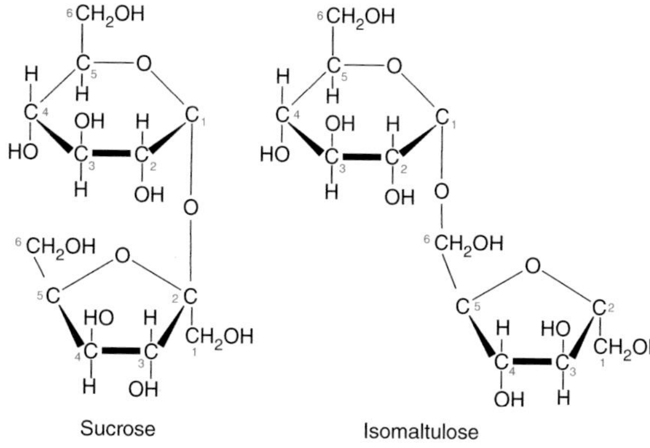

Fig. 2.6 Chemical structure of ISO and SUC. ISO and SUC are disaccharides composed of glucose and fructose. ISO can be produced from SUC by enzymatic conversion of 1,2 to 1,6 glycosidic bond

ISO appears as a white, crystalline powder similar to SUC. It has a sweetening power about half of that of SUC. Unlike SUC, ISO melts at a lower temperature (123–124 vs. 160–185 °C). ISO has been used as sweetener in Japan since 1985 (Lina et al. 2002). It has been accepted as a novel food in the European Union (EU Commision 2005) and is generally recognized as safe by the Food and Drug Administration (FDA) in the United States (FDA 2006).

2.3.3.1 Digestion

As with other disaccharides, ISO is hydrolyzed to its monosaccharide compounds in the small intestine. The membranes of intestinal epithelial cells contain brush-border enzymes, which degrade ISO by cleaving the α-1,6 glycosidic bonds between glucose and fructose molecules. One of the brush-border enzymes that catalyzes the hydrolysis of ISO is the sucrase-isomaltase complex. This enzyme consists of two polypeptide chains (sucrase and isomaltase), which are highly distributed in the proximal jejunum (Goodman 2010). Results of studies using human and rat intestinal homogenates as well as purified rat intestinal sucrase-isomaltase complex indicate that ISO is mainly hydrolyzed by the active site of isomaltase subunit (Dahlqvist et al. 1963; Goda and Hosoya 1983; Goda et al. 1988). In addition, ISO is hydrolyzed by another brush-border enzyme, i.e., glucoamylase (Dahlqvist et al. 1961), however only a small amount (Goda et al. 1988; Günther and Heymann 1998). This enzyme has its highest activity in the proximal ileum (Goodman 2010).

The hydrolysis rate of ISO is slower compared with other sugars such as SUC, maltose, or isomaltose. Studies using homogenate and brush-border membrane of rat jejunum have demonstrated that ISO is degraded at a rate approximately 20–25 % of that of SUC, 2–5 % of that of maltose (Tsuji et al. 1986; Goda et al. 1988), and approximately 10 % of that of isomaltose (Tsuji et al. 1986). This is also consistent with the finding of a study using pig intestinal mucosa homogenate (Dahlqvist et al. 1961). The maximal velocity value for the hydrolysis of ISO by brush-border membrane and purified enzyme sucrase-isomaltase complex averages 10–17 % of SUC, 13–16 % of isomaltose, and 2–7 % of maltose (Goda and Hosoya 1983; Tsuji et al. 1986; Goda et al. 1988).

Studies in healthy humans indicate that ISO is completely digested in the small intestine. Dietary carbohydrates that are nondigestible pass into the large intestine, where they are fermented by the colonic microflora with subsequent formation of short chain fatty acids (e.g., acetate, propionate, and butyrate), hydrogen, and carbon dioxide. However, no changes in the fecal microflora, fecal pH, and water contents were observed following daily administration of 24 g ISO tablet for 10 days in healthy volunteers (Kashimura et al. 1990). Further, no significant increase in breath hydrogen levels was seen within 9 h following a bolus ingestion of 10 g ISO which was administered after daily ingestion of 5 g ISO for 12 days in healthy subjects. Also, no significant differences were apparent in the serum concentrations of acetate and propionate as well as gastrointestinal symptoms such as diarrhea, flatulence, or abdominal pain (Tamura et al. 2004). Consistently, a more recent study in rats also demonstrated no significant increase in the breath hydrogen excretion over a 7-h period following oral administration of ISO at a dose of 2 g/kg (Tonouchi et al. 2011). Altogether, those findings suggest that ISO is not fermented in the colon. A study in healthy ileostomy subjects who had undergone colectomy found that approximately 95.5 % and 98.8 % of ingested ISO was digested within a 8-h period, respectively in the form of beverage and solid-liquid meal (Holub et al. 2010). Thus, irrespective of food matrix and food consistency, ISO is virtually completely hydrolyzed.

2.3.3.2 Absorption

After digestion, both monosaccharides glucose and fructose are taken up by the intestinal absorptive cells (enterocytes) via specific transport proteins. Glucose is absorbed by Na^+-dependent active transport through the sodium-dependent glucose transporter 1 (SGLT1). In the presence of Na^+, glucose molecule binds to SGLT1 with high affinity. If the intracellular Na^+ concentration is low ($\sim$10 mM), Na^+ dissociates from its binding site in SGLT1, causing a decrease in SGLT1 affinity for glucose, and glucose is then released into the cellular cytoplasm. Fructose, on the other hand, is absorbed by a diffusion mechanism using facilitative-diffusion GLUT. Among the 5 types of GLUTs, GLUT5 is able to transport fructose into the enterocytes. Subsequently, glucose and fructose move through the cytosol, exit the enterocytes via the basolateral membrane, and cross into the blood capillary using the facilitated diffusion transporter GLUT2 (Goodman 2010).

The absorption of ISO is delayed as a result of its slower hydrolyzing rate. This is confirmed in rat studies through measurement of transmural potential difference evoked by Na^+-dependent active transport of glucose (Tsuji et al. 1986; Goda et al. 1988). The potential differences evoked by ISO as well as other sugars were found to be well correlated with their hydrolyzing activities (Tsuji et al. 1986), suggesting that digestion and absorption of sugars are dependent on their digestibility by the intestinal membrane enzymes. Further, the results in human subjects with ileostomies indicate that ISO is completely absorbed; the amounts of the absorbed ISO over a period of 8 h averaged approximately 93.6–96.1 % (Holub et al. 2010). Thus, ISO is slowly but completely absorbed.

2.3.3.3 Metabolism and Excretion

Upon absorption, glucose and fructose molecules are transported via the portal vein to the liver, where they are metabolized by the well-characterized carbohydrate metabolic pathways and subsequently distributed to all the tissues. As expected, increases in plasma glucose and insulin levels were observed following oral administration of ISO in dogs and rats (Kawai et al. 1986; Tonouchi et al. 2011). The delayed absorption of ISO results in attenuated increases in glucose and insulin levels compared with SUC. As demonstrated in rats, postprandial blood glucose and plasma insulin levels were lower during the first hour following an oral ISO load at a dose of 2 g/kg body weight compared with those following a SUC load, and remained elevated 2 h thereafter (Tonouchi et al. 2011). Further, a study in rats that were adapted to high ISO or high SUC diets for 2–3 weeks demonstrated that plasma concentrations of glucose, fructose, and insulin were lower following a 3 g bolus of ISO compared with those observed after SUC bolus within a 3-h period (Häberer et al. 2009).

Results in animal and human studies indicate that only a small portion of ISO is metabolized when administered through parenteral route. After intravenous administration of ISO to dogs at a dose of 2 g, approximately 83 % of the dose was excreted in the urine during 24 h (Hall and Batt 1996). Similarly, more than 88 % of ISO was detected intact in the urine following intravenous administration of 0.5 g in healthy adults (Menzies 1974). These findings highlight the importance of digestion of ISO by the gastrointestinal tract. Thus, through the alimentary tract both monosaccharides are made available for metabolism.

ISO has been reported to be excreted only in small amounts in human and animal studies. Up to 4.5 % of ISO was detected in the fluid homogenate samples of healthy ileostomy subjects within 8 h following a 50 g load, indicating a low excretion (Holub et al. 2010). In rats administered with [^{14}C]ISO via oral gavage, approximately 2.5 % and 3.6 % of the total amounts of radioactivity were found respectively in the faeces and urine, whereas over 50 % of the radioactivity was identified in the expired air within 72 h following administration of single doses up to 0.5 g/kg body weight (Macdonald and Daniel 1983). The urinary and fecal excretion of radioactivity after ISO load was comparable to that observed after SUC load, indicating that a large amount of administered ISO undergoes metabolism as that applied for SUC. Further, the tracer findings may suggest that the majority of ISO disposal could be via the oxidative glycolytic pathway because more than half of the administered radioactivity was eliminated through the respiratory tract as carbon dioxide.

2.3.3.4 Human Studies on Glucose Metabolism

To provide an overview of ISO studies in human subjects, a database search in PubMed, Cochrane Library, and Web of Science was conducted in March 2014. The keyword "isomaltulose or palatinose" was used. No language specification or years of publication was chosen. The results of the database search are summarized in Fig. 2.7. A total of 435 abstracts was found. By using the filter search for

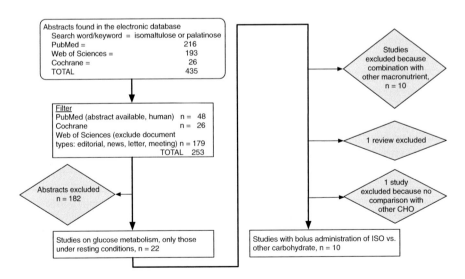

Fig. 2.7 Flow diagram of systematic review of ISO studies. *CHO* carbohydrate, *n* number

"abstract and humans" in PubMed and excluding the document types "editorial, news, letter or meeting" in Web of Science, the search was limited to 253 abstracts. After focusing on the topic of glucose metabolism and considering for overlapping publications, 22 abstracts were identified to be potentially interesting and were reviewed. Studies with bolus administration of ISO compared with other carbohydrates were included. Studies with ISO consumption combined with meal ingestion or other macronutrients were excluded in order to eliminate other potential nutrient interactions that may influence glucose metabolism. Ten published studies fulfilled the inclusion criteria. A summary description of these studies is presented in Tables 2.2, 2.3, and 2.4 respectively in healthy, impaired glucose-tolerant, and diabetic subjects.

Seven of 10 studies evaluated postprandial plasma glucose and insulin concentrations in healthy subjects over a period of 1.5–3 h (Table 2.2). The ingested dose ranged mostly between 50 and 75 g. Overall, postprandial glucose levels were lower after ingestion of ISO compared with administration of SUC, noticeably during the first hour following consumption. Peak glucose levels were approximately 20–35 % lower after 50 or 75 g load of ISO versus SUC (Kawai et al. 1985, 1989; Liao et al. 2001; van Can et al. 2009; Holub et al. 2010; Maeda et al. 2013). In contrast, incremental area under the curve (iAUC) of plasma glucose did not differ between ISO and SUC as reported in two studies (Macdonald and Daniel 1983; van Can et al. 2009). In all studies, plasma insulin levels and iAUCs or AUCs were significantly lower after ISO versus SUC bolus, approximately 50 % and 30 % lower for peak insulin level and iAUC of ISO, respectively. One study found significant higher GLP-1 levels after ISO ingestion compared with SUC (Maeda et al. 2013).

Among the 10 studies, only one study assessed the effects of ISO post-ingestion on glucose metabolism compared with SUC in impaired glucose-tolerant subjects (Table 2.3). In this study, postprandial glucose and insulin levels were lower during the first hour of a 75 g ISO load (van Can et al. 2012). When iAUCs of glucose and insulin were compared during the 3-h period, a significant difference was found for insulin but not for glucose.

Two studies examining postprandial glucose and insulin levels in healthy subjects also included measurement in T2DM patients (Kawai et al. 1989; Liao et al. 2001) (Table 2.4). In both studies, postprandial glucose and insulin levels were lower following administration of 50 and 75 g of ISO versus SUC. The cumulative increase of glucose and insulin levels as well as the AUCs of glucose and insulin were overall lower during a 3-h period after the ISO load. In addition, two other studies from the same workgroup found an attenuated postprandial increase of glucose levels following ISO bolus compared with dextrose in T1DM subjects prior to exercise (West et al. 2011; Bracken et al. 2012). In both studies, the T1DM subjects reduced their normal rapid-acting insulin dose by 50–75 % before administration of the carbohydrates. Blood lactate levels increased greater with ingestion of ISO than with dextrose.

Table 2.2 Studies with bolus ingestion of ISO in healthy subjects

Reference	n	Intervention	Design	t	Parameter	Results
Studies in healthy subjects						
Macdonald and Daniel (1983)	10	0.25, 0.5, 0.75, and 1 g/kg BW of ISO or SUC	Randomized, crossover	1.5 h	• Glucose, fructose, and insulin levels	ISO<SUC (0.5 h)
					• Glucose iAUC	ns
					• Fructose and insulin iAUCs	ISO<SUC
Kawai et al. (1985)	8	50 g ISO or SUC	Randomized, crossover	2 h	• Glucose and insulin levels	ISO<SUC (0.25–1 h)
					• Peak glucose level	ISO<SUC (~23 %)
					• Peak insulin level	ISO<SUC (~54 %)
					• Total Δ glucose and insulin levels	ISO<SUC (~50 %)
Kawai et al. (1989)	10	50 g ISO or SUC	Randomized, crossover	2 h	• Glucose and insulin levels	ISO<SUC (0.5–1 h)
					• Peak glucose level	ISO<SUC (~20 %)
					• Peak insulin level	ISO<SUC (~44 %)
					• Total Δ glucose levels	ISO<SUC (~35 %)
					• Total Δ insulin levels	ISO<SUC (~52 %)
Liao et al. (2001)	10	75 g ISO or SUC	Crossover	3 h	• Peak glucose level	ISO<SUC (~35 %)
					• Peak insulin and C-peptide levels	ISO<SUC
					• Glucose, insulin, and C-peptide AUCs	ISO<SUC
Holub et al. (2010)	10	50 g ISO or SUC	Randomized, crossover, double-blind	3 h	• Glucose and insulin levels	ISO<SUC (0.25–1 h)
					• Peak glucose level	ISO<SUC (~20 %)
					• Peak insulin level	ISO<SUC (~50 %)
					• Glucose and insulin iAUCs	ISO<SUC (~35 %)
Maeda et al. (2013)	10	50 g ISO or SUC	Crossover, double-blind	3 h	• Glucose levels	ISO<SUC (0.25–1 h)
					• Insulin levels	ISO<SUC (0.25–0.5 h)
					• Peak glucose level	ISO<SUC (~25 %)
					• Total GIP levels	ISO<SUC (0.25–1 h)
					• Total and active GLP-1 levels	ISO>SUC (1.5 h)
Studies in healthy overweight subjects						
van Can et al. (2009)[a]	10	Breakfast with 75 g ISO or SUC	Randomized, crossover, single-blind	3 h	• Peak glucose level	ISO<SUC (~20 %)
					• Peak insulin level	ISO<SUC (~50 %)
					• Glucose iAUC	ns
					• Insulin iAUC	ISO<SUC (~30 %)
					• Ghrelin level	ISO<SUC (3 h)

Values in parenthesis are the time periods in which significant differences were found or the percentage differences between ISO and SUC. *BW* body weight, *n* number of subjects, *ns* non-significant, *t* duration, Δ increase over baseline

[a]This study consisted of a breakfast and a lunch meal. Because the lunch meal contained other macronutrients, only data at breakfast were considered here

Table 2.3 Studies with bolus ingestion of ISO in impaired glucose-tolerant subjects

Reference	n	Intervention	Design	t	Parameter	Results
van Can et al. (2012)[a]	10	Breakfast with 75 g ISO or SUC	Randomized, crossover, single-blind	3 h	• Glucose and insulin levels	ISO<SUC (0.5–1 h)
					• Peak glucose level	ISO<SUC (~17 %)
					• Peak insulin level	ISO<SUC (~40 %)
					• Total glucose iAUC	ns
					• Total insulin iAUC	ISO<SUC (~21 %)

Values in parenthesis are the time periods in which significant differences were found or the percentage differences between ISO and SUC. *n* number of subjects, *ns* non-significant, *t* duration
[a]This study consisted of a breakfast and a lunch meal. Because the lunch meal contained other macronutrients, only data at breakfast were considered here

Table 2.4 Studies with bolus ingestion of ISO in diabetic subjects

Reference	n	Intervention	Design	t	Parameter	Results
Studies in T1DM subjects						
West et al. (2011)[a]	8	75 g ISO or DEX[b]	Randomized, crossover	2 h	• Δ Glucose levels	ISO<DEX (0.5–2 h)
					• Peak Δ glucose level	ISO<DEX (~50 %)
					• Lactate levels	ISO>DEX (0.5–2 h)
Bracken et al. (2012)[a]	7	0.6 g/kg BM (~40 g) ISO or DEX[c]	Randomized, crossover	2 h	• Δ Glucose levels	ISO<DEX (0.5–1.5 h)
					• Peak Δ glucose level	ISO<DEX (~45 %)
					• Lactate levels	ISO>DEX (0.5–2 h)
Studies in T2DM subjects						
Kawai et al. (1989)	10	50 g ISO or SUC	Randomized, crossover	3 h	• Glucose and insulin levels	ISO<SUC (0.5–1.5 h)
					• Peak glucose level	ISO<SUC (~18 %)
					• Peak insulin level	ISO<SUC (~26 %)
					• Total Δ glucose and insulin levels	ISO<SUC (~30 %)
Liao et al. (2001)	10	75 g ISO or SUC	Crossover	3 h	• Peak glucose level	ISO<SUC (~30 %)
					• Peak insulin and C-peptide levels	ISO<SUC
					• Glucose, insulin, and C-peptide AUCs	ISO<SUC

Values in parenthesis are the time periods in which significant differences were found or the percentage differences between ISO and SUC. *BM* body mass, *DEX* dextrose, *n* number of subjects, *t* duration, Δ increase over baseline
[a]This study was performed under resting and subsequently under exercise conditions. Only data at resting were considered here
[b]Prior to ingestion, subjects reduced their normal rapid-acting insulin dose by 75 %
[c]Prior to ingestion, subjects reduced their normal rapid-acting insulin dose by 50 %

2.3.4   Co-administration of Carbohydrate and Protein

Dietary macronutrients carbohydrates, proteins, and fats are usually ingested in a complex food matrix rather than in their pure form. As a consequence, interactions between the macronutrients may influence glycemic response. Understanding nutrient interdependencies is therefore important. For instance, food proteins alone already stimulate insulin release, whereas a notably additive effect on insulin release is triggered by the combined uptake of proteins with carbohydrates (Gannon et al. 1988, 1992; van Loon et al. 2003; Manders et al. 2005, 2006). However, controversial results have been reported in regard to the effects of insulin stimulation on glucose homeostasis in T2DM subjects. Some studies have shown a reduction of postprandial glucose response when proteins and carbohydrates were taken up in combination compared to the uptake of carbohydrates alone (Gannon et al. 1988; Manders et al. 2005, 2006). Other studies did not confirm this effect (Gannon et al. 1992; van Loon et al. 2003). In this context, various proteins were studied and it remains to be clarified whether they have comparable and consistent effects on postprandial glucose. Interestingly, animal studies indicate that a protein mixture of whey and soy has potent postprandial glucose-attenuating and insulin-stimulating effects. These were also observed when the proteins were simultaneously ingested with a slow release carbohydrate (Hageman et al. 2008).

2.4   Objectives

The objective of the work was to assess the effects of ISO consumption on postprandial glucose metabolism compared with SUC in T2DM subjects, particularly to determine the magnitude and the underlying mechanisms involved in the regulation of postprandial glucose flux. It sought to answer the question whether ISO could improve the abnormality of postprandial glucose homeostasis and be effective for glycemic control compared with SUC in T2DM.

The specific objectives were:

- to assess the kinetics of postprandial glucose absorption following a bolus ingestion of ISO compared with SUC
- to evaluate whether the secretion of incretin hormone GLP-1 can be enhanced by intake of ISO compared with SUC
- to determine the kinetic parameters involved in the regulation of glucose homeostasis, including systemic glucose appearance, oral glucose appearance, EGP, SGU, and systemic glucose disappearance
- to examine the effect of ISO consumption on insulin sensitivity compared with SUC in T2DM
- to investigate whether the co-administration of ISO either with an isonitrogenous mixture of whey and soy proteins, or with casein proteins, could increase insulin, thereby reducing postprandial glucose responses compared to ingestion of ISO alone in T2DM.

Chapter 3
Methods

Two studies were conducted in order to assess postprandial glucose metabolism after administration of ISO or SUC and after a combined load of ISO with different proteins in T2DM subjects. In the first study (ISO/SUC-Clamp Study), postprandial glucose metabolism including mechanism and effects on glucose kinetics was assessed using a double-tracer technique, which combines a hyperinsulinemic-euglycemic clamp with an oral ISO or SUC load (Ang and Linn 2014). In the second study (ISO-Protein Study), postprandial glucose and insulin responses were investigated using oral administration of ISO alone or in combination with proteins (Ang et al. 2012). Both studies were approved by the Ethics Committee of the Faculty of Medicine at Justus Liebig University (Giessen, Germany) under the registration number 99/02. The studies were registered at ClinicalTrials.gov as NCT01070238 and conducted under the supervision of Prof. Dr. Thomas Linn. All procedures involving human subjects were carried out according to the guidelines set out in the Declaration of Helsinki.

3.1 Subjects

All subjects were men and women diagnosed with T2DM according to WHO or ADA criteria. Subjects were recruited from outpatient clinics at Medical Clinic and Policlinic 3 of Justus Liebig University (Giessen, Germany). They were included in the studies if they fulfilled the following criteria:

- Diagnosis of T2DM for >1 year
- Adults between 18 and 75 years of age
- BMI between 18 and 40 kg/m^2
- HbA$_{1c}$ <64 mmol/mol (8 %) with fasting blood glucose level <7.8 mmol/L (140 mg/dL)
- On stable antidiabetic treatment for ≥2 months prior to studies

M. Ang, *Metabolic Response of Slowly Absorbed Carbohydrates in Type 2 Diabetes Mellitus*, SpringerBriefs in Systems Biology, DOI 10.1007/978-3-319-27898-8_3

Before the studies began, all subjects underwent a detailed medical history, a physical examination, and clinical laboratory tests, which included a complete blood count, blood chemistry analysis, urinalysis, pregnancy test in premenopausal women, and electrocardiography. Subjects with the following characteristics were excluded from the studies:

- T1DM
- Unstable or untreated proliferative retinopathy
- Clinically significant nephropathy, neuropathy, hepatic diseases, and heart failure
- Uncontrolled hypertension
- Systemic treatment with corticosteroids
- Pregnancy
- Insulin treatment

Subjects were informed about the studies and the potential risks. They were requested to withdraw from any medication 3 days prior to the start of the studies. Written informed consent was obtained from all subjects.

3.2   Study Design

3.2.1   Isomaltulose/Sucrose-Clamp Study

This study had a randomized, double-blind, crossover design with an interval of at least one week between two experiments. Subjects were randomly assigned to either receive ISO or SUC using computer-generated random numbers by GraphPad QuickCalcs (GraphPad Software, La Jolla, CA, USA). Coded and numbered containers with ISO or SUC were used to implement random allocation and blindness. Subjects, care providers, and outcome assessors were blinded to the interventions.

3.2.1.1   General Methodology

Each experiment consisted of a 3-h pre-ingestion phase and a 4-h postprandial phase. The pre-ingestion phase served as a preparation period, in which a euglycemic-hyperinsulinemic clamp was performed. Subsequently, it was followed by an oral administration of labeled ISO or SUC, and the postprandial responses were measured for 4 h.

Hyperinsulinemic-Euglycemic Clamp The glucose clamp experiment was first introduced by Reubin Andres in 1966 and later was developed by him and his colleagues into hyperinsulinemic-euglycemic clamp (DeFronzo et al. 1979). It has become a widely accepted method for measuring insulin sensitivity in clinical research (Muniyappa et al. 2008). This approach utilizes insulin that is usually administered as a prime-constant or constant infusion for increasing plasma insulin concentration to a new level (hyperinsulinemia). This causes an increase in systemic glucose R_d and a suppression of EGP, which result in a rapid decrease of plasma glucose level. To prevent the subjects from developing hypoglycemia, plasma glucose concentrations are kept in the normal range (euglycemia) by infusing exogenous glucose at a variable rate.

After several hours of continuous insulin infusion, steady-state conditions are achieved for plasma insulin, plasma glucose, and GINF rates. Often, it is assumed that data obtained at the end of a 2-h hyperinsulinemic-euglycemic clamp experiment represent the steady-state values. As shown in Fig. 3.1, steady-state plasma glucose, plasma insulin, and GINF rates can be defined as the average values during the relatively constant period, e.g., between 90 and 120 min. Under these conditions with a concomitant complete suppression of EGP, GINF rates should equal glucose uptake by insulin-sensitive tissues. Thus, the GINF value reflects a direct measure of tissue insulin sensitivity. In healthy subjects, EGP can be completely suppressed by infusing insulin at a rate less than 1 mU/kg/min for 2 h (Rizza et al. 1981; Soop et al. 2000). However, EGP may not be totally inhibited in insulin-resistant diabetic subjects. In such cases, systemic glucose R_d may be underestimated; it is necessary to use labeled isotopes to assess EGP and correct it from the estimated GINF.

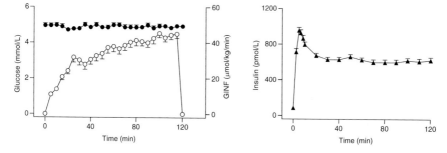

Fig. 3.1 Steady-state plasma glucose and insulin concentrations as well as GINF rates during a 2-h hyperinsulinemic-euglycemic clamp in 11 healthy subjects. Data are mean ± SEM. A 10-min primed insulin infusion was initiated, resulting in an overshoot in plasma insulin concentration (*black triangle*), followed by a declining and subsequent constant infusion rate. Plasma glucose concentrations (*black circle*) were maintained within the euglycemic range by the gradual increase in GINF rates (*white circle*) (Graphs adapted from DeFronzo et al. 1979)

Double-Isotope Technique In metabolic research, isotope techniques are often applied to determine kinetic characteristic of a substrate. Isotopes can be combined with the hyperinsulinemic-euglycemic clamp to accurately assess the dynamic aspect of glucose metabolism. The term "isotopes" refers to all forms of a chemical element having an equal number of protons but different atomic mass. There are stable and radioactive isotopes; the latter ones have unstable mass and emit radiation. Unlike radioisotopes, stable isotopes do not produce ionizing radiation, therefore they are safe and non-toxic for studies in children and adults (Rennie 1999). The less abundant stable isotopes, such as ^2H, ^{13}C, ^{15}N, and ^{18}O, are

commonly used in the metabolic kinetic studies. Due to the different masses
between stable isotopes and their siblings, they can be easily distinguished between
each other. The abundances of these isotopes are shown in Table 3.1.

Table 3.1 Stable isotopes commonly used in metabolic studies

Element	Most abundant	Natural abundance (%)	Less abundant	Natural abundance (%)
H	^{1}H	99.985	^{2}H	0.015
C	^{12}C	98.89	^{13}C	1.11
N	^{14}N	99.63	^{15}N	0.37
O	^{16}O	99.76	^{17}O	0.037
			^{18}O	0.204

Adapted from Rennie (1999) and Sakurai et al. (2000)

The basic method of stable isotope techniques involves the dilution principle
(Rennie 1999). According to the concept of mass conservation, the atomic mass of
a stable isotope must remain constant over time, as it can neither be created nor
destroyed. From this, it follows that adding an isotope as a tracer into a biological
system will not change its total mass, including that which is metabolized, exhaled,
or excreted. Thus, when a known amount of tracer n/v is added and mixed into a
pool V, measuring its dilution in the pool will allow one to calculate the size of the
pool. The concentration of tracer c in the pool equals

$$c = \frac{n}{V + v} \tag{3.1}$$

where n is the amount of tracer (mg) in volume v (mL) and V is the volume of the
pool (mL). Solving Eq. 3.1 for pool volume V, one obtains

$$V = \frac{n}{c} - v \tag{3.2}$$

The above dilution principle applies to static pools, for example in the measurement
of body composition (Rennie 1999).

In the context of human metabolism, in which the description of substrate's
movements in a compartment is essential, the dynamic tracer dilution is used. The
principle of a dynamic pool is illustrated in Fig. 3.2. Usually, an isotope tracer is
administered into a body as a prime-continuous or continuous infusion to follow a
substrate (tracee). The ideal tracer should be identical to the tracee but less abundant
than its most naturally occurring forms. To prevent significant alteration in the
tracee pool size, the tracer is infused in trace amounts (Vella and Rizza 2009). After
several hours of infusion (approximately 2 h for glucose), the tracer will equilibrate
throughout the body and reach the isotopic plateau value; this condition is called
steady-state stability or known as steady state. Under this condition, tracer infusion
rate F equals its disposal rate f; hence, tracer mass q is constant (Cobelli et al. 1992).

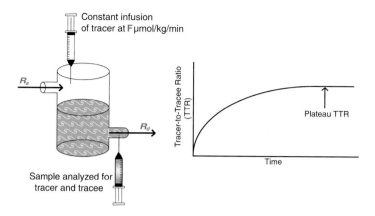

Fig. 3.2 Dynamic dilution principle applied in the metabolic studies. A tracer is administered at a constant rate to follow a tracee in a body pool. Samples are taken at several time points to determine the tracee R_a. After achieving an isotopic plateau value, which is marked by the constant value of TTR, tracee R_a is equal to the tracer infusion rate F divided by the TTR value (Figure adapted from Rennie 1999)

The mass balance equation for the tracer is

$$\frac{dq}{dt} = F - f \qquad (3.3)$$

Because a steady-state condition is assumed, tracee R_a should equal tracee R_d. Analog to Eq. 3.3, the mass balance equation for tracee Q is

$$\frac{dQ}{dt} = R_a - R_d = 0 \qquad (3.4)$$

The tracer and tracee are assumed to be indistinguishable, which implies that the probability that a particle leaves the system is a tracer equals the probability that a particle in the system is a tracer. It can be written as

$$\frac{f}{R_d + f} = \frac{q}{Q + q} \qquad (3.5)$$

From the mass conservation concept, the tracer infusion rate should equal its disposal rate. Thus,

$$F = f \qquad (3.6)$$

Substituting Eq. 3.6 for f with the Eq. 3.5 becomes

$$F = f = \frac{R_d + f}{Q + q} q \tag{3.7}$$

Solving Eq. 3.7 for R_d, one obtains

$$R_d = \frac{F}{q/Q} = R_a \tag{3.8}$$

Thus, under steady-state condition, tracee R_a equals ratio of the tracer infusion rate F and the quotient between tracer and tracee masses q/Q (Cobelli et al. 1992). In the following, this quotient is expressed as tracer-to-tracee ratio (TTR). Equation 3.8 can then be re-written as

$$R_a = \frac{F}{TTR} \tag{3.9}$$

where R_a is the tracee rate of appearance (μmol/kg/min) and F is the tracer infusion rate (μmol/kg/min). By frequent sampling, e.g., every 5 or 15 min, the steady-state condition can be identified by the relatively constant TTR values. Perturbation by food, starvation, injury or drugs can convert the steady-state condition into non-steady state. Estimation of substrate kinetics under non-steady-state condition usually involves determination of substrate R_a and R_d, which are calculated using mathematical models.

The commonly used stable isotopes in glucose kinetic analysis are [2-^{2}H]glucose, [3-^{2}H]glucose, [6,6-^{2}H$_2$]glucose, [1-^{13}C]glucose, and [U-^{13}C]glucose (Sakurai et al. 2000). [6,6-^{2}H$_2$]glucose has been generally used as a non-recycling tracer, since there is almost no chance that both ^{2}H will recycle back into the C-6 position of glucose after glycolysis and gluconeogenesis. [1-^{13}C]glucose or [U-^{13}C]glucose can also be considered as non-recycling tracers, provided that the recycling of ^{13}C during gluconeogenesis is taken into account. This can be accurately quantified by selectively monitoring specific ionized glucose molecules to differentiate the tracer from labeled glucose derived from it by using gas chromatography mass spectrometry (GCMS) (Coggan 1999).

Steele et al. (1968) introduced a double-isotope tracer technique; one tracer is infused intravenously, and the other one is ingested. This approach enables differentiation between endogenous and exogenous derived glucose. The intravenously infused tracer measures systemic glucose R_a including EGP and oral glucose released into the systemic circulation, whereas the ingested tracer is used to trace oral ingested glucose R_a. EGP is calculated by subtracting systemic glucose R_a from the ingested glucose R_a. In the ISO/SUC-Clamp Study, the dual-isotope technique was applied in combination with the hyperinsulinemic-euglycemic clamp to assess postprandial glucose kinetics following ISO or SUC load. [6,6-^{2}H$_2$]glucose was intravenously administered for measuring systemic glucose R_a, additionally [1-^{13}C]ISO or [1-^{13}C]SUC was orally ingested. ISO and SUC were labeled with ^{13}C at C-1 position in the glucose molecules. The chemical structure of the labeled ISO or SUC is depicted in Fig. 3.3. [6,6-^{2}H$_2$]glucose is a glucose molecule labeled with two ^{2}H atoms at C-6 position. ^{2}H isotope is usually called deuterium, which has an

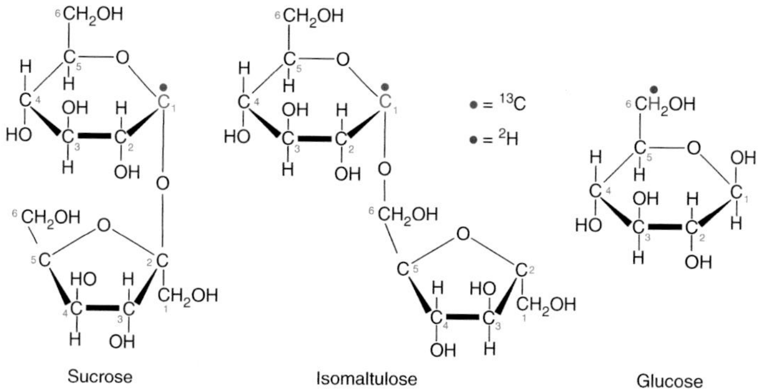

Fig. 3.3 Chemical structure of $[1\text{-}^{13}]$SUC, $[1\text{-}^{13}]$ISO, and $[6,6\text{-}^2H_2]$glucose

atomic mass of 2. By contrast, the natural most abundant isotope of this element has a mass of 1 (^1H). The position of the labeled $[6,6\text{-}^2H_2]$glucose is also illustrated in Fig. 3.3.

3.2.1.2   Experimental Procedure

The experimental procedure is depicted in Fig. 3.4 and detailed in this section. After an overnight fasting (-180 min), baseline blood samples were drawn from an antecubital vein of each subject. The hyperinsulinemic-euglycemic clamp was initiated with an infusion of $[6,6\text{-}^2H_2]$glucose through a cannulated vein on the opposite arm. This infusion was prepared by dissolving 1800 mg of $[6,6\text{-}^2H_2]$glucose in 50 mL isotonic saline (0.9 % NaCl) in a perfusor syringe. A primed dose of 8 mL (288 mg) was used to reduce the time of $[6,6\text{-}^2H_2]$glucose for reaching equilibrium and followed by a constant infusion at a rate of 3 mg/min for determining systemic glucose R_a.

Parallel to $[6,6\text{-}^2H_2]$glucose infusion, a continuous insulin infusion was administered. This was prepared by adding 20 units of insulin to 50 mL isotonic saline in a perfusor syringe. Baseline plasma insulin was gradually raised to a new steady-state level, which resulted in an increase of systemic glucose R_d and a suppression of EGP. To allow a measurement of dynamic EGP after oral ISO or SUC load, the insulin infusion was set at a rate of 0.8 mU/kg/min (Rizza et al. 1981; Campbell et al. 1988). Under this condition, plasma glucose was reduced to euglycemia (5 mmol/L) and maintained at that level by monitoring blood glucose concentrations every 5 min and adjusting infusion rate of a 20 % glucose solution. This variable GINF was enriched with $[6,6\text{-}^2H_2]$glucose to minimize changes in plasma TTR (Finegood et al. 1987). The amount of $[6,6\text{-}^2H_2]$glucose added to the GINF was calculated to approximate steady-state plasma TTR. Under steady-state condition, systemic glucose R_a equals tracer infusion rate divided by the TTR as described in Eq. 3.9.

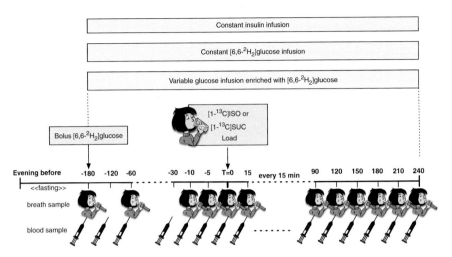

Fig. 3.4 Design of ISO/SUC-Clamp Study. After an overnight fast, a hyperinsulinemic-euglycemic clamp was initiated by continuous infusions of insulin and $[6,6-{}^2H_2]$glucose in T2DM subjects. To maintain blood glucose at euglycemic levels during hyperinsulinemic condition, an exogenous glucose infusate enriched with $[6,6-{}^2H_2]$glucose was infused. After 3-h infusions, subjects ingested randomly a drink containing either ^{13}C-enriched ISO or ^{13}C-enriched SUC. All infusions were maintained until the end of the study. Blood and breath samples were collected in parallel prior to and following ingestion according to a time-sampling procedure

Assuming that steady-state glucose R_a is $\sim 11\,\mu mol/kg/min$ in a subject with 70 kg body weight, the $[6,6-{}^2H_2]$glucose infusion with the rate of 3 mg/min ($0.24\,\mu mol/kg/min$) would result in a steady-state plasma TTR of approximately 2.16 %. To achieve an equivalent TTR, the amount of $[6,6-{}^2H_2]$glucose that needs to be added to a 500 mL of 20 % GINF solution can be simply calculated as the ratio of $[6,6-{}^2H_2]$glucose concentration to total glucose concentration.

$$TTR \approx \frac{c\,([6,6-{}^2H_2]glucose)}{c\,(20\%\ glucose)} \tag{3.10}$$

where $c\,([6,6-{}^2H_2]glucose)$ is the concentration of $[6,6-{}^2H_2]$glucose, which equals n/V. n is the amount of $[6,6-{}^2H_2]$glucose (g) and V is the volume of GINF solution (500 mL). The concentration of GINF solution is equal to 0.2 g/mL. Solving Eq. 3.10 for n, one obtains

$$n = TTR \cdot c\,(GINF) \cdot V \tag{3.11}$$

Substituting all variables yields a n-value of 2.16 g. Thus, 2.16 g of $[6,6-{}^2H_2]$glucose needs to be diluted in the GINF solution in order to obtain the approximated steady-state plasma TTR. To extrapolate the calculated amount of $[6,6-{}^2H_2]$glucose for all subjects, a mean body weight of 75 kg is used, multiplying with the required amount of ~ 2 g, a general formula is developed. Therefore, the amount of $[6,6-{}^2H_2]$glucose added to the GINF solution equals 150 g divided by kg body weight of each subject.

An additional 0.9 % isotonic saline was infused at a constant rate of 10 mL/h to maintain fluid and electrolyte balance.

After isotopic and substrate equilibrium had been achieved (0 min), the subjects received either 1 g/kg body weight of ISO or SUC in a random order. ISO and SUC were obtained in powder form from Numico Research (Wageningen, the Netherlands), dissolved in 200 mL water, and ingested within 5 min. They were enriched with 10 % [1-^{13}C]ISO and [1-^{13}C]SUC (Campro Scientific, Berlin, Germany). Following consumption of the disaccharides, blood glucose levels were aimed to remain near-euglycemia by reducing the exogenous GINF to compensate for oral glucose absorption. At the end of the experimental period (240 min), the GINF rates returned to the preload levels, indicating complete absorption of glucose derived from both ISO and SUC.

Blood samples were collected prior to ingestion at -180, -120, -60, -30, -10, -5, 0 min, and after ISO or SUC ingestion at 15 to 30 min intervals. They were placed on ice and then centrifuged at 2000 g for 10 min. Plasma aliquots were analyzed for glucose and insulin concentrations. The remaining of the samples were frozen and stored at -20 °C until the analysis of plasma concentrations of C-peptide, glucagon, GLP-1, GIP, and plasma TTRs of [6,6-^{2}H$_2$]glucose and [^{13}C]glucose. In parallel, matched breath samples were collected to measure ^{13}CO$_2$ isotopic abundance. The exhaled breath ^{13}CO$_2$ was used as an indicator of glucose oxidation.

3.2.2 Isomaltulose-Protein Study

This study had a randomized, double-blind, crossover design. Subjects were randomized to either ingest ISO combined with whey and soy mixture (ratio 1:1) or ISO combined with casein by using a randomization tool from the website Randomization.com. There was a minimum interval of 3 days between the two experiments. Additionally, in a separate experiment, the subjects were asked to consume ISO alone. The experimental procedure is described below and shown in Fig. 3.5.

After an overnight fast, basal blood samples were drawn from an antecubital vein of each subject. Subsequently, the subjects ingested randomly either 50 g ISO combined with 21 g whey/soy (ISO+WS) or 50 g ISO combined with 21 g casein (ISO+C). The amino acid composition of whey/soy and casein is listed in Fig. 3.6. In another experiment, the subjects consumed only 50 g ISO. ISO, ISO+WS, and ISO+C powders were manufactured by Numico Research (Wageningen, the Netherlands). They were dissolved in water and administered as a single bolus within 5 min. Blood samples were taken at every 15 min and at a 30-min interval following 90 min ingestion until blood glucose returned to the baseline levels. Blood samples were directly centrifuged at 2000 g for 10 min. Glucose and insulin concentrations were measured in aliquots of plasma. The rest of the plasma samples were frozen and stored at -20 °C until the analysis of amino acids concentrations.

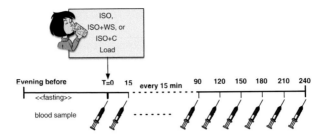

Fig. 3.5 Design of ISO-Protein Study. After an overnight fast, subjects randomly ingested a drink containing ISO+WS or a drink containing ISO+C. In another experiment, the subjects ingested a drink containing only ISO. Blood samples were drawn prior to, at a 15-min interval up to 90 min, and at a 30-min interval until the end of the experiment

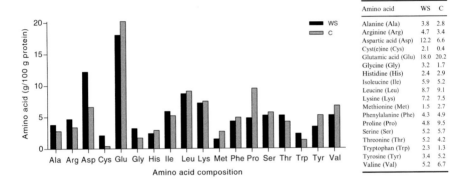

Amino acid	WS	C
Alanine (Ala)	3.8	2.8
Arginine (Arg)	4.7	3.4
Aspartic acid (Asp)	12.2	6.6
Cyst(e)ine (Cys)	2.1	0.4
Glutamic acid (Glu)	18.0	20.2
Glycine (Gly)	3.2	1.7
Histidine (His)	2.4	2.9
Isoleucine (Ile)	5.9	5.2
Leucine (Leu)	8.7	9.1
Lysine (Lys)	7.2	7.5
Methionine (Met)	1.5	2.7
Phenylalanine (Phe)	4.3	4.9
Proline (Pro)	4.8	9.5
Serine (Ser)	5.2	5.7
Threonine (Thr)	5.2	4.2
Tryptophan (Trp)	2.3	1.3
Tyrosine (Tyr)	3.4	5.2
Valine (Val)	5.2	6.7

Fig. 3.6 Amino acid composition in whey/soy (ratio 1:1) and casein. Data are in g/100 g protein. *C* casein, *WS* whey/soy

3.3 Analytics

3.3.1 Measurement of Plasma Glucose Concentrations

Plasma glucose concentrations were measured in the central laboratory of the University Hospital of Giessen and Marburg using the ADVIA 1650 Chemistry System (Bayer HealthCare, Leverkusen, Germany) based on the glucose hexokinase II method.

Principle The assay basic principle is derived from the Slein's method (Slein et al. 1950; Slein 1965; Bayer HealthCare 2006), which involves two enzymatic reactions that are determined by a change in the absorbance using a spectrophotometer. It uses enzymes hexokinase and glucose-6-phosphate dehydrogenase (G-6-PD). Hexokinase catalyzes the phosphorylation of glucose from adenosine triphosphate (ATP). The chemical reaction is irreversible, producing the substrate glucose-6-phosphate (G-6-P). This substrate is subsequently oxidized by the G-6-PD, causing the reduction of nicotinamide adenine dinucleotide (NAD) to NADH. The NADH absorbance is measured at a wavelength of 340 nm. Thus, glucose concentration is directly proportional to NADH concentration.

$$Glucose + ATP \xrightarrow{Hexokinase} G\text{-}6\text{-}P + ADP$$

$$G\text{-}6\text{-}P + NAD^+ \xrightarrow{G\text{-}6\text{-}PD} 6\text{-}Phosphogluconate + NADH + H^+$$

Procedure Plasma samples were added to the tubes containing reagent 1 composed of buffer, sodium azide, ATP, and NAD. The absorbance of each sample was immediately read at 340 nm for correcting interfering substances. Thereafter, reagent 2 was added to the tubes to initiate the enzymatic reactions of glucose. Reagent 2 containing the same compounds as reagent 1 was mixed with a buffer solution of hexokinase and G-6-PD enzymes. The absorbance was again read at 340 nm (Bayer HealthCare 2006). The difference in the absorbance unit between the first and second measurements is proportional to plasma glucose concentration, which was determined from a standard curve using known glucose concentrations.
The assay had the following characteristics (Bayer HealthCare 2006):

- Detection range of 0–700 mg/dL (0–38.9 mmol/L)
- No significant interferences ($\leq 10\%$) to 25 mg/dL of bilirubin, 500 mg/dL of hemoglobin, and 500 mg/dL of lipemia (intralipid); all at glucose concentration of 80 mg/dL
- Intra- and inter-assay coefficients of variation of $<3\%$
- Correlation of $y = 1.00x - 0.09$; $n = 35$, $r = 1.000$

3.3.2 Measurement of Plasma Hormone Concentrations

3.3.2.1 Insulin

Plasma insulin concentrations were measured in the central laboratory of the University Hospital of Giessen and Marburg using the ADVIA Centaur System (Bayer HealthCare, Leverkusen, Germany) based on a sandwich chemiluminescence immunoassay technique.

Principle The principle of the immunoassay method is based on the formation of antigen-antibody complex. In immunology, an antibody is a Y-shaped protein being produced in the presence of a foreign object called antigen. Thus, an antigen is a substance that triggers the production of antibody. An antibody can specifically recognize a particular structure on an antigen, thereby binding to each other. This binding mechanism is the basis applied in the immunoassay analysis. Usually, antigen is the substance to be detected, i.e., insulin. In this immunoassay method, two antibodies are used: a solid phase monoclonal mouse anti-insulin antibody and a monoclonal mouse anti-insulin antibody labeled with acridinium ester (Bayer HealthCare 2003). The antigen is bound between those two antibodies in a sandwich complex form. A specific substance is added to react with the labeled antibody of the complex form. It produces light emission (luminescence) which can be detected with a luminometer. The relative light units measured are directly proportional to the amount of antigen in a plasma sample and hence is a quantification of insulin concentration.

Procedure The assay procedure is performed by the ADVIA Centaur System in the sequential steps as follows. Plasma samples were first pipetted to tubes before the addition of labeled antibody. The sample mixtures were incubated at 37 °C for 5 min. By adding another antibody, the mixtures were then incubated again at 37 °C for 2.5 min. Thereafter, the unbound materials in the tubes were separated from the bound antigen-antibody complex by aspirating and washing the tubes with deionized water. Subsequently, an acid solution followed by a base one were added to the mixtures to induce the chemiluminescence reaction (Bayer HealthCare 2003). The light emissions measured are proportional to plasma insulin concentrations, which were derived from a calibration curve enclosed with the assay.

The assay had the following characteristics (Bayer HealthCare 2003):

- Detection range of 0.5–300 mU/L
- No significant cross-reactivity with proinsulin, C-peptide, gastrin-1, glucagon, and secretin
- Interferences of $\leq 6\%$ up to 125 mg/dL of hemoglobin, 1000 mg/dL of lipid, 20 mg/dL of bilirubin, and 12 g/dL of protein
- Intra- and inter-assay coefficients of variation of $<6\%$
- Dilution recovery of 84.6–109.4 %; spiking recovery of 103.8–113.8 %

3.3.2.2   C-Peptide

Plasma C-peptide concentrations were determined in the laboratory of Medical Clinic and Policlinic 3 at Justus Liebig University using an enzyme-linked immunosorbent assay (ELISA) kit from Dako (Glostrup, Denmark).

Principle The assay principle is based on the use of 2 monoclonal antibodies to form a complex with antigen in a plasma sample. One of the antibody is labeled with an enzyme. By adding an enzymatic substrate, the antigen-antibody complex can be detected through a color change. The color intensity can be read by a spectrophotometer or a plate reader and is a measure of the amount of antigen in the plasma sample, which is proportional to C-peptide concentration. The basic principle and assay procedure are shown in Fig. 3.7.

Procedure Plasma samples, calibrator, and control samples were pipetted into microwells of a microplate that was coated with a specific mouse monoclonal anti-C-peptide antibody. The antigens in the samples were then captured by the antibody, forming antigen-antibody complexes. A further anti-C-peptide monoclonal

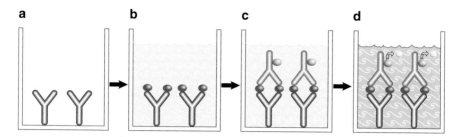

Fig. 3.7 C-peptide ELISA basic principle. (**a**) Monoclonal antibody coated microwells. (**b**) Antigen of a plasma sample binds to antibody. (**c**) A second monoclonal antibody linked to enzyme binds to the immobilized antigen. (**d**) After washing, a substrate is added and converted by the enzyme into a color product. The color intensity is proportional to the amount of antigen, i.e., plasma C-peptide concentration

antibody, which was linked to horseradish peroxidase enzyme, was added to the samples for detecting the antigens. The mixtures were incubated by placing the microplate on a plate shaker at 20–30 °C for 60 min. After incubation, the microplate was washed with buffered solution 3 times to remove the unbound enzyme-labeled antibody. Thereafter, a substrate solution containing 3,3′,5,5′-tetramethylbenzidine (TMB) was added to react with the antigen-antibody complexes. The mixtures were again incubated at 20–30 °C and shaken for 10 min. The reaction was stopped by adding sulfuric acid, producing a visual color change. The absorbance was read by an ELISA microplate reader at 450 nm. A standard curve was derived from the calibrators of known C-peptide concentrations in the assay, allowing the determination of unknown C-peptide levels in the samples (Dako 2009).

The assay had the following characteristics (Dako 2009):

- Detection limit of 0.09 ng/mL (30 pmol/L)
- No cross-reactivity with human insulin, 83–84 % cross-reactivity with intact and split proinsulin
- No interferences from lipid, bilirubin, rheumatoid factor or heterophilic antibodies; interferences with proinsulin antibody complexes
- Intra- and inter-assay coefficients of variation of <5 %
- Recovery of 93–105 %

3.3.2.3 Glucose-Dependent Insulinotropic Peptide

Plasma total GIP levels were measured in the laboratory of Medical Clinic and Policlinic 3 at Justus Liebig University using a commercial available sandwich ELISA kit of Linco Research (St. Charles, MO, USA) under the catalog number Cat. # EZHGIP-54K.

Principle The assay principle is similar to that for the measurement of C-peptide concentrations.

Procedure Plasma samples, control samples, and standard samples were added to the microwells of a microtiter plate coated with anti-GIP monoclonal antibodies and then incubated at room temperature for 1.5 h on a microtiter plate shaker. After incubation, the plate was washed 3 times with buffer solution to remove the unbound antigens in the samples. Subsequently, a second biotinylated anti-GIP polyclonal antibody was added to all microwells and incubated at room temperature for 1 h on the microplate shaker. To remove the unbound materials in the samples, the wash step was repeated 3 times. Further, an enzyme solution containing streptavidin-horseradish peroxidase conjugate was added for binding to the immobilized biotinylated antibodies and incubated at room temperature for 30 min on the plate shaker. After 3 times washing away the unbound antibody-enzyme conjugates, a TMB substrate solution was added to react with the antigen-antibody-conjugate complexes. After about 5–20 min, the reaction was stopped by adding salt acid, producing a yellow color. The color intensity was measured by a plate reader set at 450 and 590 nm (Millipore 2012b).[1] The difference in the absorbance units reflects the quantification of the enzyme activity, which is directly proportional to the unknown GIP concentrations. These were determined by using a standard curve derived from the standard samples with known GIP concentrations.

The assay had the following characteristics (Millipore 2012b):

- Detection limit of 4.2 pg/mL
- No significant cross-reactivity with glucagon, oxyntomodulin, GLP-1, and GLP-2; 100 % cross-reactivity with GIP (1-42) and GIP (3-42)
- Intra- and inter-assay coefficients of variation of <9 %
- Recovery of 81–100 %

3.3.2.4 Glucagon-Like Peptide-1

Plasma GLP-1 levels were determined in the laboratory of Medical Clinic and Policlinic 3 at Justus Liebig University on the biologically active forms, GLP-1 (7-36 amide) and GLP-1 (7-37 amide), using a commercial available ELISA kit of Linco Research (St. Charles, MO, USA) under the catalog number Cat. # EGLP-35K.

Principle The basic principle of this assay is based on a direct ELISA test (Millipore 2012a).[2] A solid-phase, immobilized monoclonal antibody binds specifically to the N-terminal region of an active GLP-1 molecule to form antibody-antigen complex. A washing step separates the unbound materials from the complex. By adding an enzyme detection conjugate and removing the unbound conjugate through a washing step, the formed antibody-enzyme conjugate can be detected

---

[1] Data sheet for Cat. # EZHGIP-54K was acquired from Millipore, formerly Linco Research.
[2] Data sheet for Cat. # EGLP-35K was acquired from Millipore, formerly Linco Research.

through chemical reaction with a specific substrate. As a result, a fluorescent outcome is produced, which directly quantifies the concentration of active GLP-1.

Procedure The assay procedure for measurement of GLP-1 is quite similar to that for the determination of GIP concentrations (Millipore 2012a). At first, a microplate coated with anti-GLP-1 monoclonal antibody was rinsed with washed buffer. Plasma samples, standards, and controls were pipetted into microwells and incubated overnight at 4 °C for 20–24 h. On the following day, the liquid was decanted and the plate was washed 5 times with the buffer solution. Immediately, a detection conjugate (anti GLP-1-alkaline phosphatase) was added for binding to the immobilized GLP-1. The mixture was incubated at room temperature for 2 h and the liquid was decanted thereafter. Next, a substrate solution containing methylumbelliferyl phosphate was added to react with the enzyme phosphatase. The mixture was incubated at room temperature for at least 20 min in the dark. After a sufficient fluorochrome was generated, a stop solution was added and the mixture was again incubated at room temperature in the dark for 5 min for terminating the enzyme activity. The relative fluorescence units were determined by a fluorescence plate reader at a wavelength of 355 or 460 nm. The measured fluorescence units are proportional to the active GLP-1 levels in the samples, which were derived from a standard curve with reference measurements of known active GLP-1 concentrations. The assay had the following characteristics (Millipore 2012a):

- Detection limit of 2 pmol/L
- No cross-reactivity with GLP-1 (9-36 amide), GLP-2, and glucagon; 100 % cross-reactivity with GLP-1 (7-36 amide) and GLP-1 (7-37 amide)
- Intra- and inter-assay coefficients of variation of $<14\%$
- Recovery of 78–100 %

3.3.2.5 Glucagon

Plasma glucagon levels were measured in the laboratory of Medical Clinic and Policlinic 3 at Justus Liebig University using a commercial radioimmunoassay (RIA) kit of Linco Research (St. Charles, MO, USA) under the catalog number Cat. # GL-32K.

Principle The basic principle of a RIA method involves application of antibody to specifically bind to antigen, which is similar to ELISA. The difference between the two methods is that RIA uses radioactive-labeled antigen whereas ELISA employs non-radioactive enzyme-linked antibody for detecting the amounts of antigens in a sample. The basic principle of a RIA assay is shown in Fig. 3.8.

In a RIA method, the radiolabeled antigen is mixed with a fixed amount of antibody to form antigen-antibody complex. Typically, about 50 % of the labeled antigen is bound to the available antibody (total binding), indicating limited binding sites. As such, when an unknown amount of unlabeled antigen in a sample is added to the mixture, both labeled and unlabeled antigens will compete for the

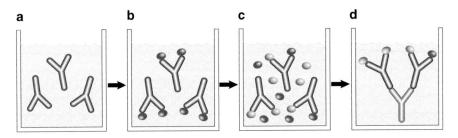

Fig. 3.8 RIA glucagon basic principle. (**a**) A mixture of antibody. (**b**) Radioactive-labeled antigen (*red dots*) binds to the antibody. (**c**) Unlabeled antigen in a plasma sample (*blue dots*) is added to the mixture and competes with the radiolabeled antigen for the limiting binding sites of antibody. The more unlabeled antigen in a plasma sample, the more labeled antigen is displaced by the unlabeled one. (**d**) A second antibody is added to the antigen-antibody complex. After washing, the complex is measured for its radioactivity

limited binding sites on the antibody. Thus, the amount of labeled antigen that binds to the antibody is correlated inversely to the amount of unlabeled antigen. This means that if the quantity of unlabeled antigen is increased, the quantity of bound labeled antigen will decrease accordingly and vice versa. After eliminating the unbound free labeled antigen, the radioactivity of the bound antigen can be measured with a gamma counter. The ratio of the bound to total bound antigen reflects the quantity of unlabeled antigen in the mixture. By using a standard curve with increasing concentrations of known amount of unlabeled antigen, the unknown concentration of unlabeled antigen can be finally determined, which quantifies the glucagon concentration in a sample.

Procedure The assay procedure took 3 days for completion (Millipore 2011).[3] On day 1, an amount of antibody was added to tubes containing only assay buffer and to tubes containing assay buffer with standards, controls, and plasma samples. The tubes were mixed and incubated at 4 °C for 20–24 h. Besides, blank tubes and other tubes comprised only assay buffer were included in the assay. On day 2, an amount of antigen i.e., glucagon labeled with ^{125}I was added to all tubes. At this time, the blank tubes contained the labeled antigen (total count (TC) tubes) and the tubes with buffer contained also the labeled antigen but without antibody (nonspecific binding (NSB) tubes). They were mixed and incubated again at 4 °C for 22–24 h. On day 3, an antibody precipitating reagent was added to all tubes, except the TC tubes. They were mixed and incubated at 4 °C for 20 min. After incubation, they were centrifuged at 4 °C for 20 min at 2,000–3,000 g. The supernatants of all tubes except those of TC tubes were decanted, and the remaining radioactive pellets were measured for their radioactivity using a gamma counter. The radioactivities of all tubes were subtracted from those of the NSB tubes, except for the TC tubes. The total amount of labeled antigen bound to antibody was calculated from the total binding tubes divided by the TC tubes. The unknown concentrations of antigen,

---

[3]Data sheet for Cat. # GL-32K was acquired from Millipore, formerly Linco Research.

hence the concentrations of glucagon in the plasma samples were determined using a standard curve, which was generated from the standard samples and calculated as the percentage ratio of standard binding to total binding.

The assay had the following characteristics (Millipore 2011):

- Detection limit of 18.5 pg/mL
- No cross-reactivity with insulin, proinsulin, C-peptide, somatostatin, and pancreatic polypeptide; <0.1 % cross-reactivity with oxyntomodulin; 100 % cross-reactivity with glucagon
- Intra- and inter-assay coefficients of variation of <7 % and 14 %, respectively
- Recovery of 96–98 %

3.3.3   Measurement of Plasma Amino Acid Concentrations

Plasma amino acid concentrations were determined in the laboratory of Numico Research (Wageningen, the Netherlands) by high-performance liquid chromatography (HPLC) using an automated pre-column derivatization with o-phthaldialdehyde and a fluorescence detector.

Principle   HPLC is a technique capable of separating and identifying several components in a sample mixture, and measuring the concentration of each component in the sample. The separation of each component relies on the interaction between a mobile phase and a stationary phase. The schematic diagram of a typical HPLC system is shown in Fig. 3.9. An HPLC equipment normally consists of a sampler, a pump, a column, and a detector. The sampler transports the sample mixture into the column, and the pump regulates the flow rate of a liquid solvent that is used as a sample diluent for carrying the sample mixture to the column. This liquid fluid is commonly composed of an aqueous organic solution (mobile phase). The column (stationary phase) is a glass or stainless steel tube packed with spherical particles or materials of differing carbon chain length. Once the sample mixture arrives at the column via the mobile phase stream, the separation of each component begins (Fig. 3.10). Each component in the mixture interacts differently with the solid adsorbent particles in the column. The stronger the interaction, the longer the component remains in the column. By contrast, it leaves the column in a relatively short time when the interaction is weak. The time at which the component emerges from the column is called the retention time, which is considered as a characteristic to identify the component. Thus, by referring to the retention time of each component in a standard sample, each composition of a sample mixture can be finally identified. The column end is connected to a detector. After exiting the column, each component is being processed by the detector. The detector generates a signal proportional to the amount of sample component, which corresponds to the sample components' concentration.

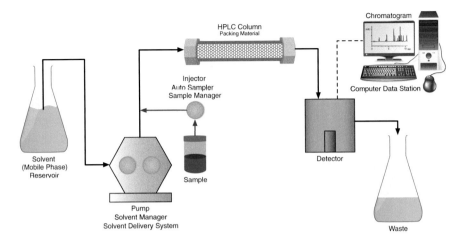

Fig. 3.9 Schematic diagram of an HPLC instrument. A sample is injected into the HPLC system. The pump regulates the flow rate of a solvent in order to carry the sample to the column. Each component of the sample interacts differently with the column, thereby being separated. After separation, it enters the detector. The computer shows an output of a chromatograph, which is called a chromatogram. The concentration of each component in the sample is proportional to the area of the chromatogram

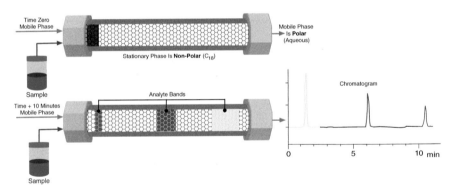

Fig. 3.10 Separation principle of each component in a sample in an HPLC column. The column shown here is made of stainless steel tube packed with MicroSpher C18 non-polar material. Thus, it has a high affinity to non-polar components. The mobile phase, usually a liquid solvent, serves as a sample carrier. When a sample reaches the column, each of the component in the sample interacts differently with the column. Less polar components (*blue band*) have naturally a higher affinity to the column and therefore remain longer in the column. On the other hand, polar components (*yellow band*) have a lesser affinity to the column and flow off readily with the polar mobile phase. As a result, the chromatogram shows the *yellow band* with a shorter retention time and the *blue* one with a longer retention time

Procedure The procedure for determination of amino acids is described as follows. Before analysis, it is necessary to hydrolyze protein and polypeptides into its individual amino acid components. For this purpose, plasma samples were deproteinized with perchloric acid. The mixtures were placed in an ice bath for 15 min and subsequently centrifuged. The remaining supernatants with free amino acids were taken for further processing in the reversed-phase HPLC apparat. Usually, the free amino acids are derivatized for analysis. A derivatization procedure is a method to modify a molecule chemically aiming at improving detection or sensitivity and reducing the time required for analysis. Thus, in the following step, the free amino acids underwent derivatization procedures before entering and being separated by the chromatographic column of the HPLC instrument. An automated pre-column derivatization method was selected by using an integrated autosampler mixing the free amino acids with o-phthaldialdehyde reagent (Fekkes et al. 1995). This reagent is soluble and stable in liquid solution with a rapid reactivity, which is suitable for the automated derivatization. The autosampler temperature was set at 20 °C and equipped with MicroSpher C18; 3.0 μm in size, length × inner diameter of 10 × 2.0 mm.

After the pre-column derivatization, the derivatized amino acids passed through the column by means of aqueous organic streams with a flow rate of 1.5 mL/min. The temperature of the column oven was set at 30 °C. Typically, in a reversed-phase HPLC, the column is packed with MicroSpher C18 material, 3.0 μm in size with a length of 100 mm and an inner diameter of 4.6 mm. This column possesses a high affinity to non-polar components. As such, the retention time is longer for amino acid components which are less polar, while polar amino acid components elute more readily in the analysis (Fig. 3.10). Measurements were made using a fluorescence detector at an excitation wavelength of 340 nm and an emission wavelength of 455 nm. To allocate each amino acid to its retention time, standard samples were included in the analysis. The concentrations of each amino acid were determined from standard curves of standard samples with the known amount of amino acids. The standards curves were created by plotting peak areas of each amino acid versus the known concentration of each amino acid in the standard samples. The peak areas of each amino acid are determined from the corresponding areas of chromatogram, i.e., a visual output of chromatography in form of a Gaussian curve (bell curve).

3.3.4 Measurement of Stable Isotope Ratios

Stable isotope ratios in plasma and breath samples were measured at the Clinical Research Center of Rochester University (NY, USA) using GCMS and isotope ratio mass spectrometry (IRMS), respectively.

3.3.4.1 Plasma [6,6-^{2}H$_2$]glucose and [^{13}C]glucose Tracer-to-Tracee Ratios

Plasma TTRs of [6,6-^{2}H$_2$]glucose and [^{13}C]glucose were measured simultaneously on the glucose acetate boronate derivative using GCMS by selectively monitoring ions of the mass-to-charge ratio (m/z) 299/297 and 298/297, respectively and corrected for natural abundance and mass isotopomer effects using calibration curves.

Principle GCMS is a combination of two instrumental techniques, namely gas chromatograph (GC) and mass spectrometer (MS) that are used to simultaneously identify and quantify different compounds in a sample. The GC performs the separation of each compound, which is further processed and characterized by the MS by measuring the mass of the compounds. Usually, the GCMS instrument comprises of a GC system with the following components: a gas supplier, a sample injector, an oven with capillary column, and a detector with the MS system (Fig. 3.11). The coupling technique of GCMS enhances the sensitivity to detect substrates qualitatively and quantitatively, even in trace amounts.

Similar to HPLC, the GC utilizes a chromatographic technique to separate each compound in a sample. When a sample is introduced by injection into the GCMS system, it is vaporized and the resulting gas sample is transferred by an inert gas as the mobile phase to the stationary phase of a GC column. The gas mixture will then be separated according to its affinity for the mobile or stationary phase. The separation principle is similar to that of the HPLC. Components with a higher affinity for the mobile phase than the stationary phase flow off the column more rapidly, while components with a higher affinity for the stationary phase are retained

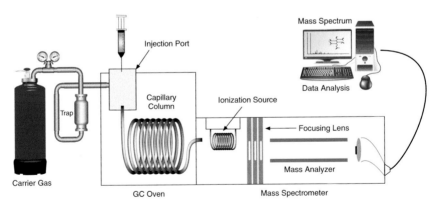

Fig. 3.11 Schematic diagram of a GCMS instrument. GCMS consists of a GC and a MS detector. A sample is injected into the system, subsequently is vaporized, and the resulting gas is carried by an inert gas to the GC capillary column. After separation of each component in a sample, it reaches the MS where the component of interest or the analyte molecule is ionized into charged molecules or ions. Typically, in a selected ion monitoring mode, only ions at a certain m/z are monitored and pass through the detector. The computer shows a mass spectrum of the monitored ions with their relative abundances

in the column, therefore are slowly eluted from the column. The GC and HPLC basic principles are similar; however, they are different in the sample types and in the composition of mobile and stationary phases. These differences are summarized in Table 3.2.

Table 3.2 Comparison of GC and HPLC

Parameter	GC	HPLC
Sample type	Gas samples (e.g., odorous samples such as petrochemicals, perfumes, and thinner) and samples that vaporize at high temperature (e.g., high molecular weight compounds can be analyzed after pyrolysis)	Liquid samples (e.g., a wide range of liquid substances) and solvent-soluble solid samples (e.g., high molecular weight compounds can be analyzed if dissolvable in liquid solvent)
Mobile phase	Gas (e.g., helium, nitrogen, hydrogen)	Liquid (e.g., mixture of water and organic solvent such as acetonitrile or methanol)
Stationary phase	Solid/liquid	Solid/liquid

After separation and elution of the individual compounds from the GC column, they pass through an interface and enter into the MS. The MS system consists of 3 important parts: an ionization chamber, an ion mass analyzer, and a detector. In the ionization chamber, each compound undergoes ionization; the most common one is the electron impact ionization. Here, the compounds are beamed with a stream of electrons. Upon collision with the electrons, the compounds are turned into charged ions. These are usually intact molecules with a single positive charge, called molecular ions ($M^{+\bullet}$).

$$M + e^- \rightarrow M^{+\bullet} + 2e^-$$

The symbol $M^{+\bullet}$ indicates molecular radical cations. Due to their instability, they can be broken into smaller pieces or fragments. The fragments can be charged or remain uncharged. Only charged compounds or fragments can pass through the ion analyzer. The charged ions have a definite mass. The mass of the fragment divided by the charge is called the *m/z*. Since most fragments have a charge of +1, the *m/z* usually represents the molecular weight of the fragment.

Most commonly, GCMS instruments use a quadrupole mass analyzer to filter ions of interest. The quadrupole consists of 4 electromagnetic rods that employ opposite voltages to control ion movements. Only ions with a defined *m/z* will reach the detector, whereas those with higher or lower *m/z* are absorbed by the rods (Patterson 1997). In a selected ion monitoring (SIM) mode, the quadrupole is set to sort ions in a limited *m/z* range for passing through the detector. The detector generates electrical signals that are proportional to the number of detected ions. The computer software shows a mass spectrum; a bar graph with x-axis representing the *m/z* ratios and y-axis representing the relative abundance of each ion. The ion abundance of a molecule at a specific mass is used for quantifying stable isotopes.

TTR for Stable Isotope Quantification TTR is an expression for the dilution of isotope tracer in a tracee system, referring to the ratio of tracer mass divided by tracee mass. The quantification of tracer dilution using radioisotopes is straightforward, since tracer and tracee are distinguishable. The tracer mass is directly proportional to the energy emitted by the radiolabeled isotope, commonly expressed as disintegrations per minute. When stable isotopes are used as tracers, the mass ratio of 2 isotopic species is measured by the MS system, which determines the ratio of the masses of labeled and unlabeled species. The unlabeled species refer to the naturally most abundant isotope, whereas the labeled species refer to the naturally less abundant one. Because stable isotopes occur naturally in every molecule (Table 3.1), they are present to some extent in the tracee and tracer. As shown in Fig. 3.12, the stable isotope tracer is basically composed of the artificially labeled species with a lower amount of the unlabeled ones, and the tracee contains the unlabeled naturally occurring isotopes (i.e., most abundant and less abundant isotopes) (Cobelli et al. 1992). For example, when a tracer labeled with ^{13}C is infused into a tracee system, the sample taken for analysis comprises a mixture of total ^{12}C and ^{13}C that is derived from both the tracer and tracee. To calculate

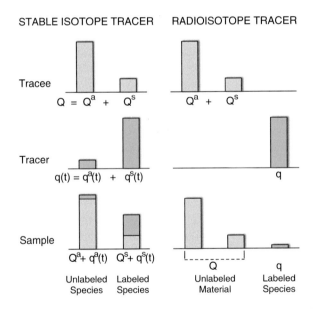

Fig. 3.12 Relative contribution of labeled and unlabeled species to a sample taken during an experiment using radioactive or stable isotope tracers. Q is tracee mass, q(t) is tracer mass, "a" superscript denotes unlabeled species, and "s" superscript denotes labeled species. Both labeled and unlabeled species are readily distinguishable when they are measured in a sample taken from a radioactive tracer experiment. In contrast, a sample taken from a stable isotope experiment contains labeled and unlabeled species, which are contributed by both the tracer and tracee. The dilution of tracer in a tracee system is expressed as TTR by dividing the tracer mass to tracee mass. Thus, $TTR = q(t)/Q$ (Figure adapted from Cobelli et al. 1992)

the unknown TTR in the sample, the contribution of tracer and tracee to each of the isotopic totals must be corrected from the known isotopic characteristic of the naturally unlabeled tracee and the known isotopic composition of the labeled tracer.

Procedure Plasma samples were initially deproteinized with acetone and the mixtures were centrifuged. The supernatants containing glucose were taken and dried under vacuum condition using a speedvac. Aliquots of glucose standards with increasing enrichment levels of [6,6-^{2}H$_2$]glucose and [^{13}C]glucose were also dried under the same condition. Further, the dried samples were derivatized before injected into the GCMS. The purpose of derivatization was to increase glucose volatility and stability, making it suitable for GC analysis. The derivatization procedure was prepared according to Wiecko and Sherman (1976). First, the dried samples were mixed with a solution of pyridine and butyl boronic acid, then placed on a shaker for 2 h. Next, acetic anhydride solution was added to the mixtures and kept them at room temperature for 1 h. Under this derivatization process, glucose was converted into glucose acetate boronate (Fig. 3.13), yielding a molecule with a molecular weight of 354. Thereafter, the samples were evaporated again to dryness using the speedvac and finally reconstituted in acetonitrile for GCMS analysis.

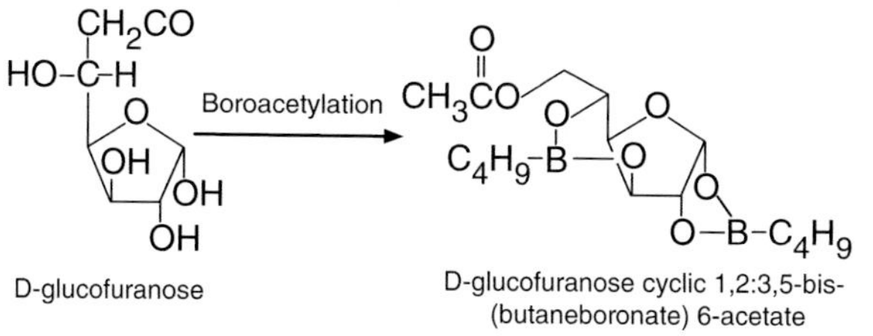

Fig. 3.13 Derivatization of glucose using boroacetylation. Glucose reacts with butyl boronic acid and acetic anhydride to form glucose acetate boronate

The enrichment of the glucose acetate boronate derivative was determined by GCMS with the SIM mode at m/z 297, 298, and 299. Briefly, after the glucose acetate boronate derivative was separated and eluted from the GC capillary column, it was ionized by the electron impact ionization method, and the molecular ions at m/z 354 were formed. The second most abundant ions in the mass spectrum are those at m/z 297, formed by expulsion of a butyl radical (C$_4$H$_9$) and charge stabilization via interaction of acetate with the boron atom of the 3,5-cyclic boronate (Fig. 3.14) (Wiecko and Sherman 1976). These ions, also called the [M-57]$^+$ ions at m/z 297 and 299, were monitored for unlabeled glucose and [6,6-^{2}H$_2$]glucose, respectively. Simultaneously, the [M-57]$^+$ ions at m/z 297 and 298 were monitored for unlabeled glucose and [^{13}C]glucose, respectively. The peak abundance ratios of ions at m/z 299/297 and 298/297 representing the ratios of ^{2}H/^{1}H and ^{13}C/^{12}C were used for calculations. Standard curves of known [6,6-^{2}H$_2$]glucose and [^{13}C]glucose enrichments were generated from each set of samples. From these ratios, the baseline-corrected and standard curve-corrected glucose enrichments were computed as TTR.

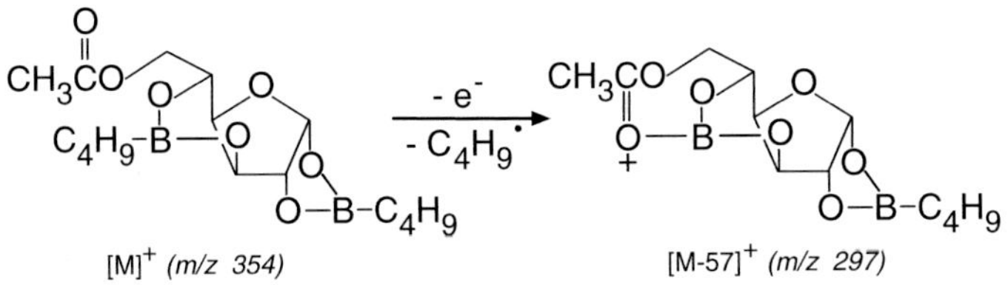

Fig. 3.14 Formation of glucose acetate boronate into fragment ion at *m/z* 297. When a glucose acetate boronate molecule is ionized, an intact molecular ion at *m/z* 354 is produced. Expulsion of a butyl radical can cause the molecular ion to break into fragment ion, yielding an ion at *m/z* 297 (Figure adapted from Wiecko and Sherman 1976)

3.3.4.2 Breath $^{13}CO_2$

Breath $^{13}CO_2$ was determined using IRMS by measuring the ratio of $^{13}C/^{12}C$ and comparing it to the international standard Pee Dee Belemnite (PDB) for carbon. Breath $^{13}CO_2$ deriving from ^{13}C-enriched carbohydrates, such as [1-^{13}C]ISO or [1-^{13}C]SUC, is used as an indicator for substrate oxidation. The pathway of substrate metabolism is shown below. After ingestion, ISO or SUC is hydrolyzed into its monosaccharides glucose and fructose. Glucose molecule reacts with oxygen to form carbon dioxide, which is then used as an indicator for the oxidative glycolysis process.

$$[1\text{-}^{13}C_{12}]H_{22}O_{11} + H_2O \longrightarrow [1\text{-}^{13}C_6]H_{12}O_6 + 6O_2 \longrightarrow 6^{13}CO_2 + 6H_2O$$

$\quad$ [^{13}C]ISO/[^{13}C]SUC $\qquad$ Water $\qquad\qquad$ [^{13}C]glucose $\qquad$ Oxygen $\qquad$ [^{13}C]Carbon dioxide $\quad$ Water

Principle An IRMS is an instrument used for measuring stable isotope ratios of particular elements, such as C, N, O, H, and S. The IRMS works on a basic principle similar to GCMS. It can differentiate the different masses of isotopes and determine the ratios between isotope pairs.

$\quad$ Generally, an IRMS consists of an inlet system, an ionizer, an ion mass analyzer, and an ion detector. The inlet system is the place where pure gases samples (e.g., CO_2, N_2, H_2, and SO_2) or biological samples are introduced. There are two types of inlet system: a continuous-flow and a dual inlet. In a continuous-flow inlet, the biological samples are first brought by gas carrier to a combustion chamber where the gas conversion of the samples takes place. In this chamber, a GC may be configured for separating different gas products. The resulting gases can then be used in the MS system for further analysis. Typically, a reference gas with known isotopic ratio is included in the analysis, which can be introduced into the system before or after the samples, or after a series of sample measurements. In a dual inlet, the system switches alternately between a pure gas sample and a reference gas that enter in turn into the MS. Both gas samples are brought by capillary tubes to a system of valves that controls the switch. The capillaries are equipped with crimps to enable regulation of gas flow (Brand 2004).

After the gas samples reach the MS, they undergo the same procedure as that in a typical MS system. The gas samples are initially ionized by a beam of electrons in the ionization chamber, and the resulting ionized gases are then separated in an electromagnetic area according to their masses. Finally, the ions reach detector that generates electrical signals proportional to the amount of ions hitting the detector. An IRMS is typically equipped with multiple detectors. Several ions can be caught and simultaneously detected by multiple Faraday cups, thereby allowing the simultaneous detection of multiple isotopes. Thus, the relative proportions of different isotopes can be calculated.

Procedure Breath samples and standard gas samples were injected into an IRMS with an auto injector and the isotopic ratio of $^{13}C/^{12}C$ was measured. The results were expressed as delta (δ) values. These are ratios that relate the carbon isotopic composition of a sample to that of a standard sample, usually the carbon from PDB limestone. The so called $\delta^{13}C_{PDB}$ is expressed as per mil (‰).

Delta values are said to be either heavier (enriched) or lighter (depleted) than a standard. For example, if a sample is said to have a delta value of $+5$ ‰ $\delta^{13}C$ then it is 5 parts in 1000 enriched in ^{13}C compared with the standard. If it has a delta value of -5 ‰ $\delta^{13}C$ then it is 5 parts in 1000 depleted in ^{13}C.

3.4 Calculations

3.4.1 Plasma Glucose Turnover

Stable isotope data, which are expressed as TTR, are used in the mathematical modeling for calculation of glucose kinetics. Models for calculating glucose turnover use the known tracer infusion rate and the measured tracer and tracee concentrations to estimate the unknown R_a and R_d of tracee at different time points. Equilibrium was achieved before administration of ISO or SUC in the ISO/SUC-Clamp Study (-10, -5, and 0 min), therefore glucose turnover rates were calculated using a derivation of Steele's steady-state equation (Steele et al. 1956) as described by Finegood et al. (1987). Following administration of the disaccharides, glucose turnover rates were calculated under non-steady-state condition as follows.

Prior to the calculation of glucose kinetics, a 5-min interpolation was performed on the TTR and glucose concentration data. The interpolation was necessary in order to obtain tracer and tracee data at equally spaced time intervals. Subsequently, the data were calculated as average values at 15 min interval and smoothed using OOPSEG, a program developed for smoothing tracer and tracee data based on optimal segments approach (Finegood and Bergman 1983; Bradley et al. 1993). The OOPSEG program minimizes measurement error and calculates a smooth curve closely to the original data points. It uses iterative methods to find line segments of noisy data, thereby minimizing changes in slope and deviation of the generated smoothed points. Thus, smoothing makes the calculation of glucose kinetics more accurate.

After data interpolation and smoothing, the postprandial glucose turnover was subsequently calculated under non-steady-state condition using Steele's 1-CM (Steele 1959) and Mari's 2-CM (Mari 1992) combined with Finegood's procedure to account for the amount of [6,6-^{2}H$_2$]glucose added to the exogenous GINF (Finegood et al. 1987). The models for estimation of non-steady glucose turnover rates are based on the following assumptions (Radziuk et al. 1978):

- Tracer's physical and chemical behavior is identical to that of tracee
- Time-variation in R_a is slower than mixing times in the compartment where both tracer and tracee appear
- Sampling of concentrations and appearance of tracer and tracee are in the same well-mixed compartment

Steele's One-Compartment Model In the Steele's model (Steele 1959; Cobelli et al. 1987), the tracee R_a with respect to time or $R_a(t)$ is estimated as follows

$$R_a(t) = \frac{R_a^*}{a(t)} - V_S \frac{C(t)\dot{a}(t)}{a(t)} \qquad (3.12)$$

where R_a^* is the tracer infusion rate given as a constant infusion, $C(t)$ is the tracee concentration at time t, $a(t)$ is the TTR at time t which equals the ratio of tracer-to-tracee concentration at time t or defined as $C^*(t)/C(t)$, and $\dot{a}(t)$ is the TTR derivative at time t and is equal to the ratio $da(t)/dt$. The parameter V_S represents the Steele's glucose distribution volume. In the body, glucose is distributed in a physically complex fashion in defined regions, such as in vascular, interstitial, and intracellular fluids. As a result of flows and diffusion, glucose is nonuniformly distributed throughout the body. Its distribution volume is usually assumed to be approximately 25 % of body weight or assumed to equal 200 mL/kg. As shown in Fig. 3.15, the Steele's model for determining the tracee R_a involves a single compartment with a constant volume V_S and a time-varying elimination rate $k_{01}(t)$. In this compartment analysis, it is assumed that rapid changes in the TTR and

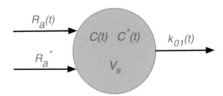

Fig. 3.15 Steele's 1-CM for calculation of glucose kinetics under non-steady-state condition. Glucose is distributed in the compartment throughout a volume V_S, which equals a constant pool fraction of the total distribution volume of glucose ($V_S = pV_T$). The pool fraction p is used to account for the nonuniform mixing of glucose in the compartment. $C(t)$ = tracee concentration, $C^*(t)$ = tracer concentration, $k_{01}(t)$ = fractional tracer disappearance rate, $R_a(t)$ = tracee rate of appearance, R_a^* = tracer rate of appearance, V_S = Steele's volume (Figure adapted from Steele 1959 and Cobelli et al. 1987)

concentration of glucose do not occur uniformly within the glucose pool. To account for the nonuniform mixing, a pool fraction was used as a correction factor (Debodo et al. 1963). Thus, V_S equals a pool fraction p of the total glucose distribution volume V_T or is expressed as $V_S = pV_T$.

Mari's Two-Compartment Model In later years, a 2-CM for calculation of glucose kinetics was proposed by Steele himself (Steele et al. 1974) and others, such as Radziuk et al. (1978), Cobelli et al. (1987), and Mari (1992) in order to eliminate the uncertainty associated with nonuniformity of glucose distribution in the pool fraction approach. The Mari's model for determining the tracee R_a in the non-steady state is equivalent to the 2-CM proposed by Steele et al. (1974) and to one of the 2-CM proposed by Radziuk et al. (1978). Compared to them, the Mari's approach is more practical because the calculations can be performed without using a specific or a complex computer program. The model is shown in Fig. 3.16 below. The first compartment is assumed to be the accessible compartment where both tracer and tracee appear, and therefore it is the compartment where glucose tracer and tracee concentrations are sampled and measured. The glucose distribution volume of this accessible compartment is constant, denoted by V_1. The second compartment is the unobservable compartment with volume V_2. Since changes in glucose concentrations are influenced by glucose disappearance rates, the exchange coefficients between the two compartments (k_{12} and k_{21}) are assumed constant. Glucose disappearance is assumed to take place only in the first compartment, therefore k_{02} is equal to 0 (Katz et al. 1974; Radziuk et al. 1978).

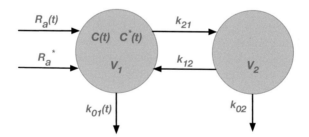

Fig. 3.16 Mari's 2-CM for calculation of glucose kinetics under non-steady-state condition. Tracer and tracee appear in the first compartment in which they are measured. $C(t)$ = tracee concentration, $C^*(t)$ = tracer concentration, k_{21} and k_{12} = intercompartmental rate constants, $k_{01}(t)$ and k_{02} = tracer disappearance rates in the respective compartment, $R_a(t)$ = tracee rate of appearance, R_a^* = tracer rate of appearance, V_1 and V_2 = glucose distribution volumes of the respective compartment (Figure adapted from Mari 1992 and Cobelli et al. 1987)

The Mari's 2-CM equation (Mari 1992) for calculating tracee R_a is

$$R_a(t) = \frac{R_a^*}{a(t)} - V_1 \frac{C(t)\dot{a}(t)}{a(t)} + R_2(t) \tag{3.13}$$

As in the Steele's 1-CM, the Mari's 2-CM equation includes in the first part a steady-state equation for the calculation of non-steady glucose turnover rates, i.e., $R_a^*/a(t)$. The second part is identical to that of Steele's, differing only in the glucose distribution volume of the compartment. The third part, denotes as $R_2(t)$, is the key method in the 2-CM and is the estimate of the structure error. The calculation of $R_2(t)$ is complex as described in Mari (1992) by a convolution integral equation. Its calculation requires the knowledge of all involved parameters, such as the volume V_2 and the rate parameters k_{01}, k_{02}, k_{21}, and k_{12}. In practical term, Mari (1992) developed an equivalent equation for estimation of $R_2(t)$, which can be implemented on a worksheet. The equation is defined as follows

$$R_2(t_k) = V_2 k_{22} \left[\frac{g^*(t_k)}{a(t_k)} - g(t_k) \right] \tag{3.14}$$

where $a(t)$ is the TTR at time t, $g(t)$ and $g^*(t)$ are the delayed glucose concentrations calculated respectively from tracee and tracer concentrations at time t, k_{22} is the sum of the rate parameters k_{12} and k_{02}, i.e., $k_{22} = k_{12} + k_{02}$, and V_2 is the glucose distribution volume of the second compartment. Total glucose distribution volume is the sum of the volume in the first and second compartments, i.e., $V_T = V_1 + V_2$. The volume V_2 is related to V_1 and the rate parameters of the model, which can be calculated as

$$V_2 = V_1 \frac{k_{12}k_{21}}{k_{22}^2} \tag{3.15}$$

The variables $R_2(t_k)$, $g(t_k)$, and $g^*(t_k)$ are $R_2(t)$, $g(t)$, and $g^*(t)$ calculated over a discrete set of time points. $R_2(t)$ is calculated iteratively from t_1 with $R_2(t_0) = 0$. Before the calculation of $R_2(t)$, the parameters $g(t_k)$ and $g^*(t_k)$ need to be first estimated. These can be calculated using the following equations

$$g(t_{k+1}) = b_1 g(t_k) + b_2 C(t_k) + b_3 C(t_{k+1}), \quad g(t_0) = C(t_0) \tag{3.16}$$

$$g^*(t_{k+1}) = b_1 g^*(t_k) + b_2 C^*(t_k) + b_3 C^*(t_{k+1}), \quad g^*(t_0) = C^*(t_0) \tag{3.17}$$

The parameters b_1, b_2, and b_3 are constants, which are solved using the following equations

$$b_1 = e^{-k_{22}T} \tag{3.18}$$

$$b_2 = \frac{1 - b_1}{k_{22}T} - b_1 \tag{3.19}$$

$$b_3 = 1 - \frac{1 - b_1}{k_{22}T} \tag{3.20}$$

T is the time interval between each data set used in the calculation. $R_2(t_k)$ is subsequently integrated in Eq. 3.13 to obtain the tracee R_a.

Finegood's Approach In the ISO/SUC-Clamp Study, two glucose tracers were applied to measure glucose kinetics. One tracer was intravenously infused and the other one was ingested. This dual-isotope method used the intravenously infused [6,6-²H₂]glucose tracer to measure total glucose R_a and the ingested [1-¹³C]ISO or [1-¹³C]SUC to estimate R_a of oral absorbed glucose deriving from ISO or SUC. The [6,6-²H₂]glucose tracer was infused both at constant and variable rates. As a variable infusion, [6,6-²H₂]glucose was added to the exogenous GINF and the GINF rate was variably adjusted according to blood glucose concentration. To account for the [6,6-²H₂]glucose tracer added to the exogenous GINF during the ISO/SUC-Clamp experiment, both 1-CM and 2-CM equations for the calculation of glucose R_a were slightly modified according to Finegood et al. (1987). Figure 3.17 shows the adjusted models, which are basically similar to the Steele's 1-CM and the Mari's 2-CM with respect to the additional variable rate of [6,6-²H₂]glucose added into the glucose pool.

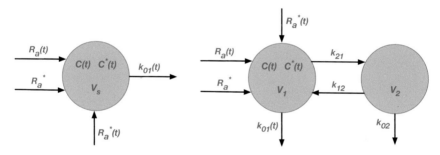

Fig. 3.17 Modified 1-CM and 2-CM for calculation of glucose kinetics under non-steady-state condition. Glucose tracer is infused both at constant and variable rates. $C(t) =$ tracee concentration, $C^*(t) =$ tracer concentration, k_{21} and $k_{12} =$ intercompartmental rate constants, $k_{01}(t)$ and $k_{02} =$ tracer disappearance rates in the respective compartment, $R_a(t) =$ tracee rate of appearance, $R_a^* =$ constant tracer rate of appearance, $R_a^*(t) =$ variable tracer rate of appearance, $V_S =$ Steele's glucose distribution volume, V_1 and $V_2 =$ glucose distribution volumes of the respective compartment

The modified 1-CM and 2-CM equations for calculating glucose R_a under non-steady-state condition become

$$R_a(t) = \frac{R_a^* + R_a^*(t)}{a(t)} - V_S \frac{C(t)\dot{a}(t)}{a(t)} \tag{3.21}$$

$$R_a(t) = \frac{R_a^* + R_a^*(t)}{a(t)} - V_1 \frac{C(t)\dot{a}(t)}{a(t)} + V_2 k_{22} \left[\frac{g^*(t_k)}{a(t_k)} - g(t_k) \right] \tag{3.22}$$

Thus, in the ISO/SUC-Clamp Study, the tracer input is the sum of constant and variable infusion rates R_a^* and $R_a^*(t)$, respectively. $R_a^*(t)$ is equal to the time-varying

GINF multiplied by the GINF enrichment or expressed as

$$R_a^*(t) = GINF(t) \cdot TTR_{GINF} \tag{3.23}$$

The calculations of each parameter of glucose kinetics are described below and used in the data analysis of ISO/SUC-Clamp Study.

3.4.1.1 Total Glucose Rates of Appearance

Total glucose R_a was estimated using the modified 1-CM and 2-CM as described in Eqs. 3.21 and 3.22, respectively. Because the measurement was based on the dilution of [6,6-^{2}H$_2$]glucose tracer, the concentration ratio of plasma [6,6-^{2}H$_2$]glucose to plasma total glucose was used in the calculation. Total glucose concentration was measured directly in the plasma sample, thereby comprising both unlabeled glucose and labeled glucose originated from the infused and ingested tracers. Thus,

$$G_T(t) = G(t) + G_{^2H}(t) + G_{^{13}C}(t)$$
$$= G(t) \left[1 + TTR_{^2H}(t) + TTR_{^{13}C}(t) \right] \tag{3.24}$$

Solving Eq. 3.24 for plasma [6,6-^{2}H$_2$]glucose concentration

$$G_{^2H}(t) = \frac{G_T(t) \cdot TTR_{^2H}(t)}{1 + TTR_{^2H}(t) + TTR_{^{13}C}(t)} \tag{3.25}$$

By substituting all relevant parameters used in Eq. 3.21, the modified 1-CM equation for calculating total glucose R_a becomes

$$R_a T(t) = \frac{F_{^2H} + F_{^2H}(t)}{G_{^2H}(t)/G_T(t)} - \frac{pVG_T(t)}{G_{^2H}(t)/G_T(t)} \cdot \frac{d\left[G_{^2H}(t)/G_T(t)\right]}{dt} \tag{3.26}$$

Similarly, considering all relevant parameters used in Eq. 3.22, the modified 2-CM equation for estimating total glucose R_a becomes

$$R_a T(t) = \frac{F_{^2H} + F_{^2H}(t)}{G_{^2H}(t)/G_T(t)} - \frac{V_1 G_T(t)}{G_{^2H}(t)/G_T(t)} \cdot \frac{d\left[G_{^2H}(t)/G_T(t)\right]}{dt}$$
$$+ V_2 k_{22} \cdot \left[\frac{g_{^2H}(t)}{G_{^2H}(t)/G_T(t)} - g_T(t) \right] \tag{3.27}$$

where

$R_a T(t)$	$=$	total glucose R_a (in μmol/kg/min)
$G_T(t)$	$=$	plasma total glucose concentration (in mmol/L)
$G_{^2H}(t)$	$=$	plasma [6,6-^{2}H$_2$]glucose concentration, determined using Eq. 3.25 (in mmol/L)

$G_{^2\mathrm{H}}(t)/G_T(t)$	$=$	ratio of tracer-to-tracee concentration, i.e., ratio of plasma [6,6-^{2}H$_2$]glucose-to-plasma total glucose concentration (dimensionless)
$F_{^2\mathrm{H}}$	$=$	constant [6,6-^{2}H$_2$]glucose infusion rate, which equals 3 mg/min or 16.67 μmol/min (in μmol/kg/min)
$F_{^2\mathrm{H}}(t)$	$=$	variable [6,6-^{2}H$_2$]glucose infusion rate, calculated using Eq. 3.23 (in μmol/kg/min)
pV	$=$	glucose distribution volume in the 1-CM, which equals a pool fraction p of the total distribution volume V (in mL/kg)
V_1	$=$	glucose distribution volume of compartment 1 in the 2-CM (in mL/kg)
V_2	$=$	glucose distribution volume of compartment 2 in the 2-CM, estimated using Eq. 3.15 (in mL/kg)
k_{22}	$=$	constant rate parameter in the 2-CM, estimated as the sum of rate parameters k_{12} and k_{02} ($k_{22} = k_{12} + k_{02}$) with k_{12} denoting transfer from compartment 2 to compartment 1 and k_{02} denoting irreversible loss from compartment 2 (in min^{-1})
$g_{^2\mathrm{H}}(t)$	$=$	delayed plasma [6,6-^{2}H$_2$]glucose concentration, calculated recursively using Eq. 3.17 (in mmol/L)
$g_T(t)$	$=$	delayed plasma total glucose concentration, calculated recursively using Eq. 3.16 (in mmol/L)

To calculate $R_aT(t)$ using the 1-CM equation, total glucose distribution volume V was initially considered to be constant and equals 200 mL/kg with a pool fraction p of 0.65 ($pV = 130$ mL/kg). Subsequently, a time-varying $pV(t)$ was calculated using the mass balance equation of [6,6-^{2}H$_2$]glucose under the following assumptions:

- Both R_a and R_d of glucose and glucose tracer take place in the accessible compartment, which have a volume pV
- R_d of [6,6-^{2}H$_2$]glucose tracer, i.e., $k_{01,^2\mathrm{H}}$, follows the profile pattern of total glucose R_d, which is defined as the ratio of glucose-to-insulin concentration multiplied by a constant factor (Best et al. 1981).

The mass balance equation of [6,6-^{2}H$_2$]glucose is

$$pV(t) \cdot dG_{^2\mathrm{H}}(t)/dt = F_{^2\mathrm{H}} + F_{^2\mathrm{H}}(t) - R_{d^2\mathrm{H}}(t)$$

$$= F_{^2\mathrm{H}} + F_{^2\mathrm{H}}(t) - k_{01,^2\mathrm{H}}(t) \cdot pV(t) \cdot G_{^2\mathrm{H}}(t) \qquad (3.28)$$

Solving Eq. 3.28 for $pV(t)$

$$pV(t) = \frac{F_{^2\mathrm{H}} + F_{^2\mathrm{H}}(t)}{dG_{^2\mathrm{H}}(t)/dt + k_{01,^2\mathrm{H}}(t) \cdot G_{^2\mathrm{H}}(t)} \qquad (3.29)$$

The estimated data were normalized by subtracting $R_aT(t)$ from the exogenous $GINF(t)$.

To calculate $R_aT(t)$ using the 2-CM equation, glucose distribution volume in the first compartment V_1 was assumed to equal $130\,\text{mL/kg}$. Glucose distribution volume in the second compartment V_2 was subsequently calculated using Eq. 3.15. For this calculation, values for the rate parameters k_{12}, k_{21}, and k_{22} were assumed to equal 0.059, 0.043, and $0.067\,\text{min}^{-1}$ (Mari 1992). The estimated data were normalized by subtracting $R_aT(t)$ from the exogenous $GINF(t)$.

During the first 3 h of insulin infusion, the exogenous GINF increased gradually. Subsequently after ISO or SUC ingestion, the GINF was reduced appropriately to compensate for the oral absorbed glucose deriving from ISO or SUC. Therefore, the above estimated data were normalized by subtracting $R_aT(t)$ from the exogenous $GINF(t)$. Thus, $R_aT(t)$ comprises both endogenous and exogenous glucose, i.e., oral glucose derived from ingestion of ISO or SUC and $[6,6-^2H_2]$glucose tracer.

3.4.1.2 Oral Glucose Rates of Appearance

After ingestion of $[1-^{13}C]ISO$ or $[1-^{13}C]SUC$, plasma $[^{13}C]$glucose TTR was used to derive plasma glucose concentration originating from oral ISO or SUC load. This was calculated as follows

$$G_{\text{ISO/SUC}}(t) = G_{^{13}C}(t)\left[1 + \frac{1}{TTR_{\text{ISO/SUC}}}\right] \tag{3.30}$$

where $G_{^{13}C}$ is the plasma $[^{13}C]$glucose concentration and $TTR_{\text{ISO/SUC}}$ is the TTR in ISO or SUC. Plasma $[^{13}C]$glucose concentration was determined analog to Eq. 3.25 for plasma $[6,6-^2H_2]$glucose concentration with $TTR_{^2H}(t)$ in place of $TTR_{^{13}C}(t)$.

$$G_{^{13}C}(t) = \frac{G_T(t)\cdot TTR_{^{13}C}(t)}{1 + TTR_{^2H}(t) + TTR_{^{13}C}(t)} \tag{3.31}$$

To estimate oral glucose R_a after ISO or SUC load, the concentration ratio of plasma $[6,6-^2H_2]$glucose to plasma glucose deriving from ISO or SUC was used in the calculation. Oral glucose R_a was calculated by applying Eq. 3.26, with $G_T(t)$ in place of plasma glucose concentration after ISO or SUC bolus ($G_{\text{ISO/SUC}}$). Substituting all relevant parameters, the modified 1-CM equation for calculating oral glucose R_a becomes

$$R_aO(t) = \frac{F_{^2H} + F_{^2H}(t)}{G_{^2H}(t)/G_{\text{ISO/SUC}}(t)} - \frac{pVG_{\text{ISO/SUC}}(t)}{G_{^2H}(t)/G_{\text{ISO/SUC}}(t)}\cdot\frac{d\,[G_{^2H}(t)/G_{\text{ISO/SUC}}(t)]}{dt} \tag{3.32}$$

Similarly, oral glucose R_a was calculated by applying Eq. 3.27, with $G_T(t)$ in place of plasma glucose concentration after ISO or SUC bolus ($G_{\text{ISO/SUC}}$). Considering all relevant parameters, the modified 2-CM equation for estimating oral glucose R_a becomes

$$R_aO(t) = \frac{F_{^2H} + F_{^2H}(t)}{G_{^2H}(t)/G_{ISO/SUC}(t)} - \frac{V_1 G_{ISO/SUC}(t)}{G_{^2H}(t)/G_{ISO/SUC}(t)} \cdot \frac{d\left[G_{^2H}(t)/G_{ISO/SUC}(t)\right]}{dt}$$

$$+ V_2 k_{22} \cdot \left[\frac{g_{^2H}(t)}{G_{^2H}(t)/G_{ISO/SUC}(t)} - g_{ISO/SUC}(t)\right] \quad (3.33)$$

where

$R_aO(t)$ = oral glucose R_a (in μmol/kg/min)

$G_{ISO/SUC}(t)$ = plasma glucose concentration deriving from ISO or SUC, calculated using Eq. 3.30 (in mmol/L)

$G_{^2H}(t)/G_{ISO/SUC}(t)$ = ratio of tracer-to-tracee concentration, i.e., ratio of plasma [6,6-^{2}H$_2$]glucose-to-plasma glucose concentration deriving from ISO or SUC (dimensionless)

$g_{ISO/SUC}(t)$ = delayed plasma glucose concentration deriving from ISO or SUC, calculated recursively using Eq. 3.16 (in mmol/L)

To calculate $R_aO(t)$ using the 1-CM and 2-CM equations, pV and V_1 were assumed to be constant and equal 130 mL/kg. All other parameters were estimated analog to those described for the calculation of $R_aT(t)$. The estimated data were normalized by subtracting $R_aO(t)$ from the decrement in $GINF(t)$.

3.4.1.3 Endogenous Glucose Production

Total glucose R_a comprises exogenous ingested glucose deriving from ISO or SUC ingestion plus the infused [6,6-^{2}H$_2$]glucose tracer as well as endogenously produced glucose. Therefore, EGP was calculated by subtracting oral glucose R_a and the total [6,6-^{2}H$_2$]glucose infusion rate from the total glucose R_a.

$$EGP(t) = R_aT(t) - R_aO(t) - [F_{^2H} + F_{^2H}(t)] \quad (3.34)$$

where $EGP(t)$ is the EGP at time t in μmol/kg/min. The above equation was used to calculate EGP in the 1-CM and 2-CM with their respective $R_aT(t)$ and $R_aO(t)$.

3.4.1.4 Total Glucose Rates of Disappearance

Total glucose R_d was calculated in the 1-CM using the formula

$$R_dT(t) = R_aT(t) - pV\frac{dG_T(t)}{dt} \quad (3.35)$$

The above equation can also be expressed as

$$R_dT(t) = [R_aO(t) + EGP(t) + F_{^2H} + F_{^2H}(t)] - pV\frac{dG_T(t)}{dt} \quad (3.36)$$

Similarly, total glucose R_d was calculated in the 2-CM using a slightly different formula

$$R_d T(t) = R_a T(t) - V_1 \frac{dG_T(t)}{dt} \tag{3.37}$$

where $R_d T(t)$ is the total glucose R_d at time t in μmol/kg/min.

The calculation procedure of $R_d T(t)$ in the 1-CM was similar to that for estimating $R_a T(t)$. The glucose distribution volume pV was initially assumed to equal 130 mL/kg and subsequently was varied according to Eq. 3.29. For the calculation of $R_d T(t)$ using the 2-CM equation, V_1 was assumed to be 130 mL/kg.

3.4.1.5   Splanchnic Glucose Uptake

Glucose uptake by the splanchnic tissues (i.e., hepatic and gastrointestinal tissues) was calculated by subtracting the cumulative amount of $R_a O$ from the amount of glucose ingested.

$$SGU = D - \int_0^{240} R_a O(t) dt \tag{3.38}$$

where

SGU	$=$	splanchnic glucose uptake (in g)
D	$=$	total amount of glucose ingested (in g)
$\int_0^{240} R_a O(t) dt$	$=$	total amount of $R_a O$ reaching the systemic circulation, calculated as AUC of $R_a O(t)$ during the 4-h postprandial period (in g)

The amount of ISO or SUC ingested was 1 g/kg body weight of each subject. Since ISO and SUC consist of 50 % glucose and 50 % fructose, the oral glucose load is equivalent to half of the amount of ISO or SUC ingested. Thus, D was assumed to equal half amount of ISO or SUC load. The total amount of $R_a O$ in g was calculated as follows. First, the $R_a O$ calculated at every time point was multiplied by time interval and body weight of each subject. Thereby, the amount of $R_a O$ in g at every time interval was obtained. Next, these amounts were integrated to finally obtain the total amount of $R_a O$. The same procedure was also applied to obtain the total amount of $R_a T$, EGP, and $R_d T$ in g.

3.4.2 Insulin Sensitivity

Insulin sensitivity was calculated under fasting and postprandial conditions using homeostasis model assessment (HOMA) and oral glucose minimal model, respectively.

3.4.2.1 Fasting Insulin Sensitivity

Fasting insulin sensitivity was calculated according to Matthews et al. (1985) based on the HOMA method. The model was developed by Matthews and his co-workers to interpret basal plasma glucose and insulin concentrations and the relationship between them. It was based on available physiological data of β-cell, hepatic, and peripheral responses to control glucose and insulin fluxes in humans and animals. Originally, it was summarized in a response-curve format that allows one to quantify both the degree of β-cell response and insulin resistance from basal plasma glucose and insulin concentrations. Thus, the relative importance of the contribution of β-cell deficiency and insulin resistance to diabetes can be estimated (Turner et al. 1979).

The central of the HOMA method is the interaction in the feedback loop between liver and β-cells. Basal insulin secretion rate is predominantly determined by plasma glucose concentration. Hepatic glucose production is influenced by both plasma insulin and glucose concentrations. In a simple method, the model of this interaction can be described in two equations.

$$I = f_1(B, G) \tag{3.39}$$

where I is the plasma insulin concentration, G is the plasma glucose concentration, and B is the fractional coefficient (or percentage) of normal β-cell capacity, and

$$G = f_2(R, I) \tag{3.40}$$

where R is the degree of insulin resistance relative to normal. The simultaneous equations can be solved to determine steady-state values of G and I as functions of the parameters B and R (Turner et al. 1982).

Figure 3.18 shows the diagram of glucose and insulin fluxes as well as their regulation in a dynamic feedback loop that has been used in the original HOMA method. Plasma glucose and insulin concentrations are defined as functions of β-cell secretion, hepatic glucose production, and glucose utilization by peripheral muscle, adipose tissue, and the nervous system. Basal plasma glucose concentration is regulated by hepatic glucose production, which is dependent on insulin. Glucose is produced in the liver, used by the brain, and taken up by the insulin-sensitive tissues including muscle and adipose cells. Hepatic glucose production will decrease as plasma glucose and insulin concentrations rise. Conversely, it will be stimulated

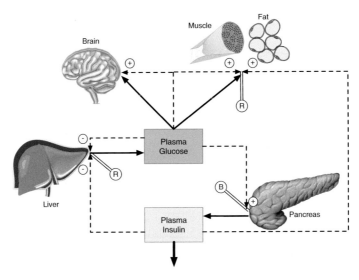

Fig. 3.18 HOMA of insulin resistance and β-cell function. Plasma glucose and insulin concentrations are functions of β-cell secretion, hepatic glucose production, and peripheral glucose utilization by muscle, adipose tissue, and the nervous system. Plasma glucose flux determines β-cell secretion from pancreas. β-cell response, in turn, regulates plasma insulin level. Insulin inhibits hepatic glucose production and accelerates glucose utilization by muscle, fat, and brain cells. Hepatic glucose production regulates plasma glucose level. Glucose per se controls its disposal into muscle, fat, and brain cells. Thus, plasma glucose and insulin concentrations are determined by their interactions in a dynamic feedback loop. *Dashed arrow* represents modulation of plasma glucose and insulin concentrations. *Solid arrow* represents glucose and insulin fluxes. B = degree of β-cell dysfunction. R = insulin resistance in liver, muscle, and fat cells. $\oplus$ = increase, $\ominus$ = decrease (Figure adapted from Turner et al. 1982)

with the increase in insulin resistance. Brain glucose uptake depends predominantly only on plasma glucose concentration. Muscle glucose uptake, by contrast, is determined by both the glucose and insulin concentrations. Its uptake will be enhanced when plasma glucose and insulin concentrations rise. On the contrary, it will decrease as insulin resistance increases (Turner et al. 1979, 1982; Hosker et al. 1985).

Plasma insulin concentration, in turn, is dependent on β-cell responses to glucose. Thus, major determinant of insulin secretion rate in the basal state is the plasma glucose concentration. Insulin is produced by β-cells and its secretion rate depends upon plasma glucose concentration. If insulin secretion by β-cells is reduced, basal plasma glucose level will increase. If β-cell response at different glucose concentrations is known, the degree of β-cell deficiency corresponding to any raised basal glucose level can be estimated. Conversely, increased insulin resistance at the liver site will enhance hepatic glucose production and plasma glucose concentration. Portal vein insulin levels will then rise until the increased glucose efflux is stabilized. Therefore, basal plasma insulin level is a function of the degree of insulin resistance (Turner et al. 1979).

In the HOMA method, defects in insulin secretion are characterized by the reduced maximal secretory response of insulin to glucose relative to normal β-cell sensitivity. Decreased β-cell function is modeled by decreasing the β-cell curve proportionally for all plasma glucose concentrations. Insulin resistance is modeled by proportionately decreasing the effect of plasma insulin concentrations at both the liver and periphery (Turner et al. 1979, 1982; Hosker et al. 1985). The resulting baseline plasma glucose and insulin concentrations are presented as a graph. The degree of insulin resistance and β-cell deficiency can be directly read off the graph illustrated in Matthews et al. (1985).

In practical terms, simple equations for determining insulin resistance and β-cells function have been proposed by Matthews et al. (1985), producing the same approximations as those described in the graph. The equations are as follows

$$HOMA\text{-}IR = \frac{G_b \cdot I_b}{22.5} \tag{3.41}$$

$$HOMA\text{-}\beta = \frac{20 \cdot I_b}{G_b - 3.5} \tag{3.42}$$

where

$HOMA\text{-}IR$	$=$	homeostasis model assessment of insulin resistance (dimensionless)
$HOMA\text{-}\beta$	$=$	homeostasis model assessment of β-cell function (in %)
G_b	$=$	fasting plasma glucose concentration (in mmol/L)
I_b	$=$	fasting plasma insulin concentration (in μU/mL)

Both equations were used to interpret basal plasma glucose and insulin concentrations. Both HOMA-IR and HOMA-β have been shown to be correlated with the euglycemic clamp test ($r = 0.88, P < 0.0001$ and $r = 0.61, P < 0.01$, respectively) (Matthews et al. 1985).

3.4.2.2 Postprandial Insulin Sensitivity

Postprandial insulin sensitivity was calculated using oral glucose minimal model proposed by Caumo et al. (2000) and Dalla Man et al. (2002). Both methods are used to derive S_I from an oral test, i.e., a meal glucose tolerance test (MGTT) or an OGTT. The model combines the classic minimal model of glucose kinetics with a mathematical description of R_aO released into the systemic circulation. The original minimal model was first developed for quantification of insulin sensitivity during an IVGTT (Bergman et al. 1979).

Bergman's Minimal Model In the Bergman's minimal model as shown in Fig. 3.19, glucose pool is illustrated as a single extracellular compartment. The amount of glucose in the pool (i.e., plasma glucose concentration) is determined by the difference between net hepatic glucose production and peripheral tissue glucose uptake. Net hepatic glucose production, in turn, depends on the difference between glucose production and glucose uptake. Glucose per se has the effects to promote its disposal into the periphery and inhibit its production, represented respectively by the rate parameters k_1 and k_5.

In this model, insulin is assumed to act from a compartment remote from plasma (Sherwin et al. 1974; Insel et al. 1975; Zeleznik and Roth 1978). This compartment could be bound insulin, insulin mediator, and/or the intracellular metabolic effects of insulin. Plasma insulin first enters the remote active insulin compartment with the rate parameter k_2, and acts to accelerate peripheral glucose disposal and inhibit net hepatic glucose production. This action, represented by the respective rate parameters k_4 and k_6, is proportional to insulin concentration in the remote compartment. The degradation rate of remote active insulin is represented as k_3, which might be related to recycling of insulin receptor, insulin mediator, and insulin induced metabolic enzymes (Bergman et al. 1979, 1985).

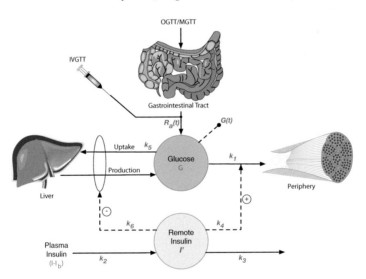

Fig. 3.19 Minimal model of glucose kinetics for estimation of insulin sensitivity from IVGTT and OGTT/MGTT. Glucose R_a is determined by the injected glucose dose in IVGTT and by the oral absorbed glucose in OGTT/MGTT. Plasma glucose is a balance of net hepatic glucose production and peripheral glucose uptake. Glucose per se inhibits hepatic production and peripheral disposal. k_1 and k_5 = rate parameters of glucose efficiency to promote disposal and inhibit production, respectively. Plasma insulin enters the remote active compartment I'. From here, it acts to inhibit hepatic glucose production and increase peripheral glucose uptake. k_2 = rate parameter of insulin distribution, binding kinetics, and insulin mediator generation. k_3 = rate parameter of insulin degradation or disappearance of effect. k_4 and k_6 = rate parameters of insulin efficiency to increase glucose disposal and inhibit glucose production, respectively (Figure adapted from Caumo et al. 2000)

The minimal model can be described in 2 equations

$$
\begin{cases}
\dot{G}(t) = -[p_1 + X(t)] \cdot G(t) + p_4 \\
\dot{X}(t) = -p_2 \cdot X(t) + p_3 \cdot I(t)
\end{cases}
\tag{3.43}
$$

The parameters p_1, p_2, p_3, and p_4 are the rate parameters, which depend on the individual k's illustrated in Fig. 3.19. The equations describe the glucose changes over time $G(t)$ and the time dependent effect of dynamic insulin response $X(t)$. $X(t)$ is a measure of insulin action to promote glucose disposal and inhibit glucose production, which is proportional to insulin concentration in the remote compartment. The first equation describes that the glucose changes are dependent on glucose per se $[p_1 G(t)]$ and on the interaction of glucose and remote insulin $[X(t)G(t)]$. Thus, if there were no dynamic insulin response, net glucose R_d would depend on p_1. Therefore, p_1 is a measure of glucose effect to enhance its own disappearance at basal insulin independent of any increase in insulin (Bergman et al. 1979, 1985). S_I derived from this model has been demonstrated to be strongly correlated with the S_I estimated from euglycemic glucose clamp ($r = 0.82$, $P < 0.005$) (Finegood et al. 1984).

Caumo's and Dalla Man's Oral Glucose Minimal Model Oral glucose minimal model proposed by Caumo et al. (2000) and Dalla Man et al. (2002) is equivalent to the minimal model developed by Bergman et al. (1979). Caumo et al. and Dalla Man et al. extended the original model by considering the absorption and appearance of oral glucose in the systemic circulation. In the IVGTT model, glucose enters directly into the systemic circulation, whereas during OGTT/MGTT glucose reaches the circulation after absorption from gastrointestinal tract and passage through the liver (Fig. 3.19). Taking $R_a O$ into account, the minimal model for oral glucose administration becomes

$$
\begin{cases}
\dot{G}(t) = -[p_1 + X(t)] \cdot G(t) + p_1 \cdot G_b + \frac{R_a O(t)}{V}; & G(0) = G_b \\
\dot{X}(t) = -p_2 \cdot X(t) + p_3 \cdot [I(t) - I_b]; & X(0) = 0
\end{cases}
\tag{3.44}
$$

where

$G(t)$	=	plasma glucose concentration with G_b denoting its basal value (in mmol/L)
$I(t)$	=	plasma insulin concentration with I_b denoting its basal value (in pmol/L)
$X(t)$	=	insulin action on glucose production and glucose disposal (in min^{-1})
$R_a O(t)$	=	plasma oral glucose R_a (in μmol/kg/min)
V	=	glucose distribution volume (in mL/kg)
p_1	=	fractional glucose effectiveness measuring glucose ability per se to promote glucose disposal and inhibit glucose production (in min^{-1})

p_2 =   constant rate of the remote insulin compartment from which insulin action is emanated (in min^{-1})

p_3 =   rate parameter governing the magnitude of insulin action (in min^{-2} per $\mu U/mL$)

From the model, S_I can be calculated as follows

$$S_I = \frac{p_3}{p_2} \qquad (3.45)$$

In term of the rate parameters of the minimal model, the above ratio can be expressed as

$$S_I = \frac{k_2(k_4 + k_6)}{k_3} \qquad (3.46)$$

The individual k's, however, cannot be directly obtained from the IVGTT or the MGTT/OGTT, but the values of p_2 and p_3 can be estimated. The S_I value increases proportionally to the rate parameters k_2, k_4, and k_6. With increased k_2, insulin will be efficiently transported into the remote compartment. With increased k_4 or k_6, insulin action will be enhanced. In contrast, when k_3 increases, more insulin will be degraded, consequently S_I will be diminished (Bergman et al. 1985).

S_I measures the overall effect of insulin to stimulate glucose disposal and inhibit glucose production. S_I is expressed as min^{-1} per $\mu U/mL$, i.e., fractional glucose disappearance per insulin concentration unit. A S_I of $0.00050\,min^{-1}$ per $\mu U/mL$ means that for each $10\,\mu U/mL$ increase in plasma insulin, there will be an increase of $0.5\,\%/min$ in fractional glucose disappearance. The fractional S_I per unit volume can be calculated as

$$S_I = \frac{p_3}{p_2} \cdot V \qquad (3.47)$$

where S_I is the insulin sensitivity index in $dL/kg \cdot min$ per $pmol/L$. S_I derived from oral glucose minimal model is strongly correlated with the S_I of tracer method ($r = 0.86$, $P < 0.0001$) (Dalla Man et al. 2004) and the S_I of euglycemic clamp experiment ($r = 0.81$, $P < 0.001$) (Dalla Man et al. 2005).

Oral glucose minimal models of Caumo et al. and Dalla Man et al. are principally similar. The difference between the both methods is the way how the S_I is calculated. Caumo et al. (2000) used an integral equation approach, whereas Dalla Man et al. (2002) used the differential equations (Eq. 3.44) to solve S_I.

In the ISO/SUC-Clamp Study, S_I was calculated using the following procedure. First, the postprandial $R_a O$ was estimated by calculating it using Eq. 3.32 for the 1-CM and Eq. 3.33 for the 2-CM. Subsequently, S_I was calculated by solving the differential equations (Eq. 3.44) and substituting the Eq. 3.45. The equation for calculating S_I becomes

$$S_I = \frac{1}{I(t) - I_b} \left[\frac{\frac{R_a O(t)}{V} + p_1 \cdot G_b - \dot{G}(t)}{G(t)} - p_1 \right] \tag{3.48}$$

V was assumed to be 200 mL/kg (Basu et al. 2003), and p_1 was assumed to be 0.014/min based on mean value reported for normal subjects (Best et al. 1996).

In the ISO-Protein Study, S_I was calculated by means of plasma glucose and insulin concentrations according to the approach of Caumo et al. (2000). The profile pattern of $R_a O$ over time is assumed to be an anticipated version of plasma glucose concentrations above their basal values. By using this assumption and integrating the oral minimal model equations, the following expression for S_I is obtained by calculating AUC of the measured plasma glucose and insulin concentrations:

$$S_I = \frac{f \cdot D_{oral} \frac{AUC[\Delta G(t)/G(t)]}{AUC[\Delta G(t)]} - S_G \cdot AUC[\Delta G(t)/G(t)]}{AUC[\Delta I(t)]} \tag{3.49}$$

When plasma glucose concentration falls below baseline, the equation becomes

$$S_I = \frac{f \cdot D_{oral} \frac{AUC[\Delta G(t)/G(t)]_0^{t_0} - AUC[\Delta G(t)/G(t)]_{t_0}^{\infty}}{AUC[\Delta G(t)]_0^{t_0} - AUC[\Delta G(t)]_{t_0}^{\infty}} - S_G \cdot AUC[\Delta G(t)/G(t)]}{AUC[\Delta I(t)]} \tag{3.50}$$

where

AUC	=	area under the curve, calculated from time 0 to the end of the test (in mg/dL·min and pmol/L·min, respectively for glucose and insulin)
S_G	=	fractional glucose effectiveness per unit distribution volume with $S_G = p_1 V$ (in mL/kg·min)
$G(t)$	=	plasma glucose concentration (in mg/dL)
$\Delta G(t)$	=	plasma glucose concentration above baseline with $\Delta G(t) = G(t) - G_b$ (in mg/dL)
$\Delta I(t)$	=	plasma insulin concentration above baseline with $\Delta I(t) = I(t) - I_b$ (in pmol/L)
f	=	fraction of ingested glucose that appears in the systemic circulation (dimensionless)
D_{oral}	=	dose of ingested carbohydrate per unit of body weight (in mg/kg)
t_0	=	time point when plasma glucose concentration crosses the baseline level, which can be determined by linearly interpolating plasma glucose concentrations that immediately precede and follow the crossing of the baseline level (in min).

The difference between Eq. 3.50 and Eq. 3.49 is that the AUCs of $\Delta G/G$ and ΔG are separately calculated in the intervals $0-t_0$ and $t_0-\infty$, and the negative AUC calculated in the interval $t_0-\infty$ is subtracted from the positive AUC calculated in the interval $0-t_0$. Prior to the calculation, the AUCs of plasma glucose and insulin concentrations were first determined by the trapezoidal method using GraphPad Prism 5 (GraphPad Software, La Jolla, CA, USA). Subsequently, the AUCs were integrated in the equation. S_G was fixed at $0.024\,dL/kg\cdot min$, and f was set at 0.8 (Caumo et al. 2000).

3.5 Statistics

3.5.1 Sample Size

For the ISO/SUC-Clamp Study, sample size calculation was based on the following assumptions:

- Mean difference in blood glucose concentration expected to be $>20\,\%$
- A standard deviation of approximately $12\,\%$ with a two-sided type I error of $5\,\%$
- A power of $>90\,\%$

Power calculation showed that at least 5 subjects were required in order to fulfill the above criteria.

For the ISO-Protein Study, sample size was calculated using the results of a previous Numico's study in 10 healthy subjects. The following assumptions were applied:

- Mean difference in glucose iAUC expected to be $>30\,\%$
- A standard deviation of approximately $55\,\%$ with a one-sided type I error of $5\,\%$
- A power of $>80\,\%$

Power calculation showed that at least 15 subjects were needed.

3.5.2 Statistical Analysis

All data are expressed as mean $\pm$ SEM. The data normality was evaluated using the Kolmogorov-Smirnov test. Comparisons between the tests at different time points were analyzed by using two-way repeated measures analysis of variance (ANOVA) with test and time as within-subject factors. If the sphericity assumption for homogeneity of variances was violated, the Greenhouse-Geisser correction was used.

For non-time-dependent variables, differences between ISO and SUC tests were compared using paired t-test or Wilcoxon signed-rank test to analyze data set with or without normality of distribution, respectively. For non-time-dependent variables, differences between ISO, ISO+WS, and ISO+C tests were compared using one-way repeated measures ANOVA or Friedman test to analyze data set with or without normality of distribution, respectively. If the test revealed significant changes and the normality assumption was not violated, Bonferroni post-hoc test was further applied for multiple comparisons between the drinks. For data set without normality of distribution, Wilcoxon signed-rank test with Bonferroni correction was chosen as post-hoc test for multiple comparisons between the drinks. Spearman's correlation coefficient (r_s) was used to assess the relationship between insulin and amino acids.

Differences within the test at different time points were analyzed by using one-way repeated measures ANOVA with Bonferroni post-hoc test for normally distributed data set. For non-normal distributed data set, Friedman test was applied, followed by Wilcoxon signed-rank test with Bonferroni correction. Analysis was performed by using SPSS 16.0 (SPSS, Chicago, IL, USA). Differences between means with an error probability of $P < 0.05$ were considered statistically significant.

Chapter 4
Results

4.1 Isomaltulose/Sucrose-Clamp Study

4.1.1 Subjects' Characteristics

Eleven T2DM subjects were included in this study. There were 5 men and 6 women aged between 40 and 65 years. Duration of diabetes was a mean of 5 years. They were obese but had good glycemic control with HbA_{1c} less than 53 mmol/mol (7 %). Three of the subjects were under dietary regimen, and the remaining 8 were treated with metformin. No other anti-diabetic medication was used. Metformin administration was stopped 3 days prior to the tests. None of the subjects had a history of unstable or untreated proliferative retinopathy, clinically significant nephropathy, neuropathy, hepatic disease, heart failure, uncontrolled hypertension, systemic treatment with corticosteroids, and insulin treatment. As determined by the HOMA approach, subjects were insulin resistant and had a β-cell function of approximately 70 %. The characteristics of these subjects are summarized in Table 4.1.

4.1.2 Plasma Glucose Concentrations

Plasma glucose concentrations were similar at baseline on both study days when ISO or SUC was ingested ($\sim$7.5 mmol/L, $P = 0.531$). During insulin infusion, basal plasma glucose concentrations decreased gradually, reaching euglycemic levels $\sim$30 min prior to ingestion of ISO or SUC (both $P < 0.01$ vs. baseline). Subsequently after administration of both disaccharides, plasma glucose concentrations climbed to peak at $\sim$80 min and returned to pre-ingestion values by $\sim$240 min.

At several time points as illustrated in Fig. 4.1, plasma glucose concentrations were lower after ISO versus SUC consumption ($P < 0.001$), with a $\sim$25 %

© The Author 2016
M. Ang, *Metabolic Response of Slowly Absorbed Carbohydrates in Type 2 Diabetes Mellitus*, SpringerBriefs in Systems Biology, DOI 10.1007/978-3-319-27898-8_4

Table 4.1 Subjects'
characteristics

Number of subjects	11
Sex (female/male)	6/5
Age (years)	53.7 ± 2.5
Body weight (kg)	90.1 ± 4.6
Height (m)	1.7 ± 0.0
BMI (kg/m^2)	31.6 ± 1.3
HbA_{1c} (mmol/mol)	49.9 ± 1.0
HbA_{1c} (%)	6.7 ± 0.1
Fasting plasma glucose level (mmol/L)[a, b]	7.5 ± 0.2
Fasting plasma insulin level (pmol/L)[a, c]	103.0 ± 14.8
HOMA-IR[d]	5.0 ± 0.8
HOMA-β (%)	73.9 ± 9.0
Diabetes duration (years)	4.6 ± 0.8
Treatment (diet/metformin)	3/8
Diet (female/male)	2/1
Metformin (female/male)	4/4

Data are mean $\pm$ SEM
[a]Values prior to the start of hyperinsulinemic-euglycemic
clamp and ingestion of ISO or SUC
[b]Measurement was in mg/dL and converted to mmol/L by
dividing by 18
[c]Measurement was in mU/L and converted to pmol/L by
multiplying by 6.945
[d]A cut-off value of >2.6 is considered insulin resistance
(Ascaso et al. 2003)

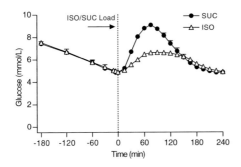

	ISO	SUC
Level (mmol/L)		
Basal, -180 min	7.5 ± 0.2	7.4 ± 0.2
Prior load, 0 min	5.0 ± 0.1	4.9 ± 0.1
Peak	6.8 ± 0.1	9.1 ± 0.1[**]
Mean, 0–240 min	5.8 ± 0.1	6.5 ± 0.1[***]
Time (min)		
Peak	83.2 ± 7.9	76.4 ± 2.4
iAUC (mmol/L·min)		
0–240 min	228.4 ± 24.5	421.7 ± 36.3[**]

Fig. 4.1 Plasma glucose concentrations prior to and after ingestion of [1-^{13}C]ISO or [1-^{13}C]SUC
in 11 T2DM subjects. A significant test effect ($P < 0.001$) and test $\times$ time interaction ($P < 0.001$)
were found. Data are mean $\pm$ SEM. ***$P < 0.001$, **$P < 0.01$ ISO vs. SUC

lower peak after the ISO load ($P = 0.003$). Consequently, mean plasma glucose
concentration was significantly lower following ISO ingestion than that following
SUC ingestion ($P < 0.001$). Plasma glucose response, iAUC, was ~45 % lower
after ISO compared with SUC intake ($P = 0.001$).

4.1.3 Plasma Hormone Concentrations

4.1.3.1 Insulin

Plasma insulin concentrations were similar at baseline between the study days on which ISO or SUC was ingested ($\sim$100 pmol/L, $P = 0.061$). During the pre-ISO or pre-SUC bolus, basal plasma insulin concentrations were increased $\sim$3.5-fold (both $P < 0.01$ vs. baseline), reaching a steady-state value $\sim$30 min before both disaccharides loads ($\sim$350 pmol/L, $P = 0.306$). As shown in Fig. 4.2, plasma insulin concentrations increased further subsequent to administration of SUC, attained a peak at $\sim$80 min, and returned thereafter to preload value by $\sim$240 min.

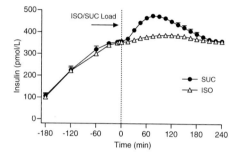

	ISO	SUC
	Level (pmol/L)	
Basal, -180 min	100. 3 ± 15.7	105. 7 ± 13.9
Prior load, 0 min	352. 4 ± 9.0	355. 8 ± 7.5
Peak	389. 5 ± 7.7	479. 8 ± 6.9 ***
Mean, 0 – 240 min	371. 6 ± 7.1	413. 3 ± 6.5 ***
	Time (min)	
Peak	113. 2 ± 8.7	79. 1 ± 5.0 *
	iAUC (pmol/L·min)	
0 – 240 min	4878. 5 ± 883.3	14685. 5 ± 1223.4 ***

Fig. 4.2 Plasma insulin concentrations prior to and after ingestion of $[1\text{-}^{13}\text{C}]$ISO or $[1\text{-}^{13}\text{C}]$SUC in 11 T2DM subjects. A significant test effect ($P < 0.001$) and test × time interaction ($P < 0.001$) were found. Data are mean ± SEM. ***$P < 0.001$, *$P < 0.05$ ISO vs. SUC

Plasma insulin concentrations increased slightly but significantly from the pre-load value following ISO ingestion ($P < 0.001$) and returned to pre-ingestion values by $\sim$240 min. Compared with SUC, plasma insulin levels were overall lower following ISO ingestion and reached a $\sim$20 % lower peak ($P < 0.001$) approximately 30 min later ($P = 0.015$). As a result, mean plasma insulin level was significantly lower after ISO ingestion than after SUC ingestion ($P < 0.001$). Accordingly, plasma insulin response was $\sim$65 % lower after administration of ISO versus SUC ($P < 0.001$).

4.1.3.2 C-Peptide

The time course of plasma C-peptide concentrations prior to and after ISO or SUC load is shown in Fig. 4.3. Baseline plasma C-peptide concentrations were identical on both study days when ISO or SUC was ingested ($\sim$1 nmol/L, $P = 0.372$). As a consequence of insulin infusion, basal plasma C-peptide levels declined similarly by 60 % during the pre-ISO or pre-SUC bolus (both $P < 0.01$ vs. baseline).

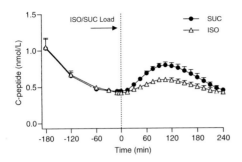

Fig. 4.3 Plasma C-peptide concentrations prior to and after ingestion of [1-^{13}C]ISO or [1-^{13}C]SUC in 11 T2DM subjects. A significant test effect ($P < 0.001$) and test × time interaction ($P < 0.001$) were found. Data are mean ± SEM. ***$P < 0.001$, **$P < 0.01$ ISO vs. SUC

Subsequently after consumption of one of the disaccharides, plasma C-peptide concentrations increased again (both $P < 0.001$ vs. preload values), reached a zenith at ~100 min, and returned to preload values by ~240 min.

At several time points, plasma C-peptide levels were lower after ISO ingestion ($P < 0.001$), with a 25 % lower peak compared with SUC intake ($P = 0.003$). Consequently, mean plasma C-peptide level was significantly lower after administration of ISO compared with SUC ($P < 0.001$). Plasma C-peptide response was accordingly ~50 % lower following ISO versus SUC bolus ($P < 0.001$).

4.1.3.3 Glucagon

Figure 4.4 shows the time course of plasma glucagon concentrations prior to and after ingestion of ISO or SUC. Plasma glucagon levels were similar at baseline on both study days when ISO or SUC was ingested (~30 pmol/L, $P = 0.371$). During insulin infusion, baseline plasma glucagon concentrations declined by ~45 % to

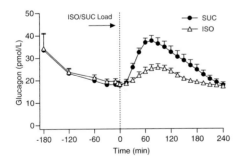

Fig. 4.4 Plasma glucagon concentrations prior to and after ingestion of [1-^{13}C]ISO or [1-^{13}C]SUC in 11 T2DM subjects. A significant test effect ($P < 0.001$) and test × time interaction ($P < 0.001$) were found. Data are mean ± SEM. ***$P < 0.001$, **$P < 0.01$ ISO vs. SUC

similar levels prior to ingestion of ISO or SUC (both $P < 0.01$ vs. baseline). Following administration of each disaccharide, plasma glucagon levels increased again from the preload values (both $P < 0.001$), reaching their zenith between 75 and 90 min and subsequently returned to preload values by $\sim$240 min. During this 4-h postprandial period, plasma glucagon levels were lower ($P < 0.001$); peak level was $\sim$30 % lower after ISO versus SUC bolus ($P = 0.003$). Thus, mean plasma glucagon concentration was consequently lower following administration of ISO as opposed to SUC ($P < 0.001$). Further, plasma glucagon response was significantly diminished by $\sim$70 % following ISO intake compared with that observed following the SUC load ($P < 0.001$).

4.1.3.4 Glucagon-Like Peptide-1

Plasma GLP-1 concentrations were similar at baseline prior to the initiation of insulin infusion on both study days when oral ISO or SUC was administered ($\sim$10 pmol/L, $P = 0.304$). During insulin infusion, plasma GLP-1 concentrations did not change significantly from the baseline levels on both ISO and SUC study days ($P = 0.185$ and $P = 0.212$, respectively), and therefore remained similar prior to ingestion of both disaccharides ($P = 0.635$). After carbohydrate intake as depicted in Fig. 4.5, plasma GLP-1 concentrations were enhanced, first ascended to a maximum at $\sim$90 min, and then decreased progressively toward preload values by $\sim$240 min. Overall, plasma GLP-1 levels were higher after ISO ingestion than after SUC administration ($P < 0.001$); the peak level was approximately 2-fold greater after ISO compared with SUC load ($P < 0.001$). Accordingly, mean plasma GLP-1 concentration was higher following ISO versus SUC ingestion ($P < 0.001$). Plasma GLP-1 responses followed a similar pattern as mean values, with a $\sim$170 % increase after the ISO bolus ($P < 0.001$).

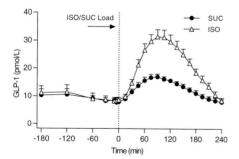

	ISO	SUC
	Level (pmol/L)	
Basal, -180 min	11.2±2.4	10.3±2.2
Prior load, 0 min	8.5±1.0	8.1±1.0
Peak	32.4±2.6	17.9±1.0***
Mean, 0–240 min	20.3±1.3	12.4±0.9***
	Time (min)	
Peak	92.7±4.4	92.7±7.0
	iAUC (pmol/L·min)	
0–240 min	3024.3±304.3	1114.2±151.5***

Fig. 4.5 Plasma GLP-1 concentrations prior to and after ingestion of [1-^{13}C]ISO or [1-^{13}C]SUC in 11 T2DM subjects. A significant test effect ($P < 0.001$) and test × time interaction ($P < 0.001$) were found. Data are mean ± SEM. ***$P < 0.001$ ISO vs. SUC

4.1.3.5 Glucose-Dependent Insulinotropic Peptide

Baseline plasma GIP concentrations did not differ from each other between the study days when oral ISO or SUC was administered ($\sim$30 pmol/L, $P = 0.753$). As with GLP-1 hormone, secretion of GIP was not affected by insulin infusion alone and remained constant on both ISO and SUC study days ($P = 0.169$ and $P = 0.067$ vs. baseline, respectively). Therefore, plasma GIP levels were similar prior to ingestion of the disaccharides ($P = 0.248$). Subsequently after SUC intake, plasma GIP levels increased from the preload value ($P < 0.001$), reached a maximum at $\sim$90 min, and then returned again to the pre-ingestion level by $\sim$240 min. A slight but significant increase in plasma GIP concentrations was observed following ISO intake ($P < 0.001$). As shown in Fig. 4.6, plasma GIP concentrations were overall lower during the 4-h postprandial period following ISO ingestion compared with those following SUC ingestion ($P < 0.001$); peak level was $\sim$22 % lower after the ISO load ($P = 0.002$). Accordingly, mean GIP concentration was lower after ISO versus SUC bolus ($P = 0.021$). Plasma GIP response was diminished by approximately 80 % following the ISO load ($P = 0.003$).

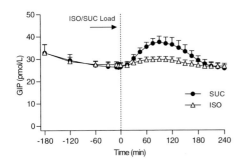

	ISO	SUC
	Level (pmol/L)	
Basal, -180 min	32.9±3.6	32.8±3.8
Prior load, 0 min	27.3±1.3	26.3±1.7
Peak	30.7±1.3	39.1±2.7**
Mean, 0–240 min	28.1±1.3	31.5±2.0*
	Time (min)	
Peak	85.9±9.5	92.7±6.7
	iAUC (pmol/L·min)	
0–240 min	287.2±80.2	1337.8±255.2**

Fig. 4.6 Plasma GIP concentrations prior to and after ingestion of [1-^{13}C]ISO or [1-^{13}C]SUC in 11 T2DM subjects. A significant test effect ($P = 0.036$) and test × time interaction ($P < 0.001$) were found. Data are mean ± SEM. **$P < 0.01$, *$P < 0.05$ ISO vs. SUC

4.1.4 Breath $^{13}CO_2$

Breath $^{13}CO_2$ excretion, an indicator for substrate oxidation, was similar at baseline on both study days prior to insulin infusion and ISO or SUC load ($P = 0.234$). As shown in Fig. 4.7, breath $^{13}CO_2$ excretion expectedly remained constant under hyperinsulinemic condition and did not differ between the days prior to ingestion of ISO or SUC ($P = 0.206$). After ISO consumption, breath $^{13}CO_2$ increased slowly up to above the pre-ingestion value and was significantly higher after 45 min of ingestion compared to the preload level ($P < 0.05$). In contrast, a significant rise was already observed after 15 min ingestion of SUC ($P < 0.05$ vs. preload value).

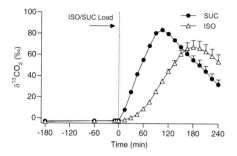

	ISO	SUC
Level (‰)		
Basal, -180 min	-3.5±0.8	-3.1±0.7
Prior load, 0 min	-2.6±0.7	-2.3±0.7
Peak	71.2±4.3	83.9±1.1**
Mean, 0–240 min	36.4±2.6	50.7±1.6***
Time (min)		
Peak	175.9±9.3	100.9±2.1**

Fig. 4.7 Breath $\delta^{13}CO_2$ excretion prior to and after ingestion of [1-^{13}C]ISO or [1-^{13}C]SUC in 11 T2DM subjects. A significant test effect ($P < 0.001$) and test × time interaction ($P < 0.001$) were found. Data are mean ± SEM. ***$P < 0.001$, **$P < 0.01$ ISO vs. SUC

As a result, peak breath $^{13}CO_2$ occurred approximately 75 min later after ingestion of ISO as opposed to SUC ($P = 0.003$), with a ~15 % lower value after the ISO bolus ($P = 0.003$). Mean breath $^{13}CO_2$ level was ~28 % lower after ISO compared with SUC intake ($P < 0.001$).

4.1.5   Glucose Infusion Rates

Glucose infusion increased slightly during insulin infusion to similar rates prior to ingestion of ISO or SUC (~14 μmol/kg/min, $P = 0.722$). After ISO or SUC load as depicted in Fig. 4.8, the exogenous GINF decreased significantly from the preload values between 30 and 45 min of post-ingestion (both $P < 0.05$) in order to compensate for the incoming oral absorbed glucose that was released into the systemic circulation. Approximately after ~90 min of both disaccharides ingestion, the GINF decreased to its nadir rate, which was lower after the SUC load than after

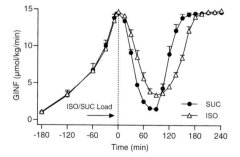

	ISO	SUC
Rate (μmol/kg/min)		
Basal, -180 min	1.1±0.2	1.0±0.2
Prior load, 0 min	14.2±0.3	14.2±0.3
Nadir	3.3±0.1	1.4±0.1***
Mean, 0–240 min	9.9±0.1	10.2±0.2
Time (min)		
Nadir	88.6±3.8	87.3±2.7

Fig. 4.8 GINF rates prior to and after ingestion of [1-^{13}C]ISO or [1-^{13}C]SUC in 11 T2DM subjects. No significant test effect ($P = 0.250$) but a significant test × time interaction ($P < 0.001$) was found. Data are mean ± SEM. ***$P < 0.001$ ISO vs. SUC

the ISO load ($P < 0.001$), and subsequently increased again to preload values. A prolonged period of time was observed in ISO for returning to the prior preload GINF compared with SUC (193.6 ± 5.5 vs. 158.2 ± 6.2 min, $P = 0.005$), indicating a slower absorption of ISO.

4.1.6   Plasma Glucose Turnover

4.1.6.1   [6,6-^{2}H$_2$]glucose and [^{13}C]glucose Tracer-to-Tracee Ratios

Plasma [6,6-^{2}H$_2$]glucose TTR increased continuously to approach a steady-state value close to 2.2 % during the 3 h after the start of the primed-constant infusion and before the administration of ISO or SUC. After ISO or SUC load, plasma [6,6-^{2}H$_2$]glucose TTR decreased to a minimum value at $\sim$75 min but was lower after the SUC bolus ($P < 0.001$), and increased again to near preload values by the end of the experiment (Fig. 4.9). Overall, the decrements in TTR were more pronounced following administration of SUC ($P < 0.001$ vs. ISO). Consequently, mean TTR value was significantly greater after ISO compared with SUC consumption ($P < 0.001$).

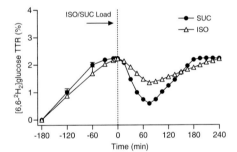

Fig. 4.9   Plasma [6,6-^{2}H$_2$]glucose TTR prior to and after ingestion of [1-^{13}C]ISO or [1-^{13}C]SUC in 11 T2DM subjects. A significant test effect ($P < 0.001$) and test × time interaction ($P < 0.001$) were found. Data are mean $\pm$ SEM. ***$P < 0.001$ ISO vs. SUC

Soon after ingestion of [1-^{13}C]ISO or [1-^{13}C]SUC, plasma [^{13}C]glucose TTR increased from zero and reached a maximum value at a later time point after the ISO load compared with the SUC load ($P = 0.014$), and then declined progressively to near zero by the end of the experiments (Fig. 4.10). Plasma [^{13}C]glucose TTRs were overall lower during the first 2 h following ISO versus SUC consumption ($P < 0.001$), with a $\sim$30 % lower peak following the ISO bolus ($P = 0.003$).

Figure 4.11 shows both plasma [6,6-^{2}H$_2$]glucose and [^{13}C]glucose concentrations that were estimated from their respective TTRs following consumption of

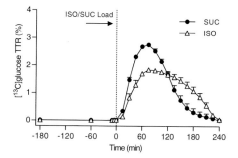

	ISO	SUC
Level (%)		
Prior load, 0 min	0.0±0.0	0.0±0.0
Peak	1.9±0.0	2.8±0.1**
Mean, 0–240 min	1.1±0.1	1.1±0.1
Time (min)		
Peak	83.2±5.1	69.5±2.3*

Fig. 4.10 Plasma [^{13}C]glucose TTR prior to and after ingestion of [1-^{13}C]ISO or [1-^{13}C]SUC in 11 T2DM subjects. No significant test effect ($P = 0.482$) but a significant test × time interaction ($P < 0.001$) were found. Data are mean ± SEM. **$P < 0.01$, *$P < 0.05$ ISO vs. SUC

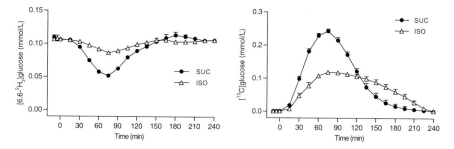

	[6,6-^{2}H$_2$]glucose		[^{13}C]glucose	
	ISO	SUC	ISO	SUC
Prior load level, 0 min (mmol/L)	0.11±0.00	0.11±0.00	0.00±0.00	0.00±0.00
Peak/nadir level (mmol/L)	0.09±0.00	0.05±0.00***	0.12±0.00	0.24±0.01***
Peak/nadir time (min)	75.0±2.0	72.3±1.8	81.8±5.1	72.3±1.8
Mean level,0–240 min (mmol/L)	0.10±0.00	0.09±0.00**	0.06±0.00	0.09±0.01***

Fig. 4.11 Plasma [6,6-^{2}H$_2$]glucose and [^{13}C]glucose concentrations after ingestion of [1-^{13}C]ISO or [1-^{13}C]SUC in 11 T2DM subjects. A significant test effect ($P = 0.002$) and test × time interaction ($P < 0.001$) were found for plasma [6,6-^{2}H$_2$]glucose concentrations. Similarly, a significant test effect ($P < 0.001$) and test × time interaction ($P < 0.001$) were found for plasma [^{13}C]glucose concentrations. Data are mean ± SEM. ***$P < 0.001$, **$P < 0.01$ ISO vs. SUC

ISO or SUC. As with plasma [6,6-^{2}H$_2$]glucose TTRs, plasma tracer concentrations decreased similarly to a minimum value at approximately 75 min after each disaccharide load but was ~45 % lower after the SUC bolus ($P < 0.001$), and increased again to a value near the preload level between 180 and 240 min. Mean [6,6-^{2}H$_2$]glucose level was significantly higher after administration of ISO as opposed to SUC ($P = 0.002$), indicating overall lower glucose turnover rates following ISO ingestion.

In contrast to the decrement of plasma $[6,6\text{-}^2H_2]$glucose concentrations following ingestion of both disaccharides, plasma $[^{13}C]$glucose concentrations increased immediately as expected, reaching a peak at approximately 75 min, and thereafter returned over time to zero by the end of the postprandial period. Peak level after the ISO bolus was half of that observed following the SUC load ($P < 0.001$). Although plasma $[^{13}C]$glucose concentrations followed a similar pattern to the plasma TTRs, the difference in the peak level between ISO and SUC was greater compared to the difference in the peak TTR ($\sim$50% vs. $\sim$30%). At several time points, plasma $[^{13}C]$glucose concentrations were lower during the first 2 h after intake of ISO compared with SUC ($P < 0.001$). Mean $[^{13}C]$glucose level during the 4-h postprandial period was significantly lower after ISO as opposed to SUC administration ($P < 0.001$). Both plasma $[6,6\text{-}^2H_2]$glucose and $[^{13}C]$glucose concentrations were used in the estimation of postprandial glucose kinetics after administration of ISO or SUC.

4.1.6.2   Total Glucose Rates of Appearance

Plasma glucose R_aT was comparable in all subjects prior to administration of each disaccharide ($\sim$9 μmol/kg/min, $P = 0.790$). As illustrated in Fig. 4.12, after ISO load, glucose R_aT rose steadily and reached a plateau at $\sim$60 min before returning toward the preload value between 210 and 240 min. After SUC load, glucose R_aT began to rise immediately, peaking at $\sim$60 min, and soon returned to the preload rate by $\sim$150 min. During the 4-h postprandial period, glucose R_aT increased significantly ($P < 0.001$) albeit to lower rates after ISO ingestion, and reached a $\sim$65% lower peak compared with SUC ($P < 0.001$). Consequently, a $\sim$35% lesser amount of total glucose appeared in the systemic circulation after ISO versus SUC administration ($P = 0.003$).

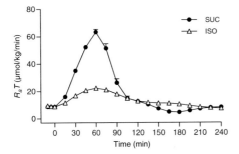

	ISO	SUC
Rate (μmol/kg/min)		
Prior load, 0 min	9.1±0.3	8.7±0.4
Peak	22.5±1.2	63.8±2.2***
Mean, 0–240 min	13.2±0.6	19.8±0.8***
Time (min)		
Peak	62.7±1.8	61.4±1.4
Total amount (g)		
0–240 min	53.5±0.8	80.1±1.3**

Fig. 4.12 Glucose R_aT after ingestion of $[1\text{-}^{13}C]$ISO or $[1\text{-}^{13}C]$SUC in 11 T2DM subjects. A significant test effect ($P < 0.001$) and test × time interaction ($P < 0.001$) were found for glucose R_aT calculated with the 2-CM approach using a constant pV of 130 mL/kg. Data are mean ± SEM. ***$P < 0.001$, **$P < 0.01$ ISO vs. SUC

4.1.6.3 Oral Glucose Rates of Appearance

Figure 4.13 shows the time course of plasma glucose R_aO after ingestion of ISO or SUC during the 4-h postprandial period. Glucose R_aO increased significantly after intake of both disaccharides ($P < 0.001$), accelerating to a maximum value at ~60 min. Thereafter, the rates decelerated over time to near zero but approximately 50 min later following ISO load compared with the SUC load ($P = 0.005$), indicating a slower ISO absorption rate. Overall, glucose R_aO values were lower after ingestion of ISO than after intake of SUC during the first 2 h. The peak rate following ISO bolus was more than half of that observed following the SUC load ($P < 0.001$). Thus, mean glucose R_aO was significantly lower after ISO consumption compared with that following SUC administration ($P < 0.001$). Accordingly, a ~25 % lesser amount of glucose derived from ISO reached the systemic circulation compared with SUC ($P < 0.001$).

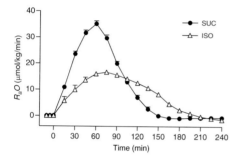

	ISO	SUC
Rate (μmol/kg/min)		
Prior load, 0 min	0.0±0.0	0.0±0.0
Peak	17.0±0.6	35.3±1.1[***]
Mean, 0–240 min	7.6±0.5	10.0±0.4[***]
Time (min)		
Peak	68.2±3.7	61.4±1.4
End of absorption	211.4±3.2	160.9±8.8[**]
Total amount (g)		
0–240 min	30.6±1.3	40.8±1.9[***]

Fig. 4.13 Glucose R_aO after ingestion of [1-^{13}C]ISO or [1-^{13}C]SUC in 11 T2DM subjects. A significant test effect ($P < 0.001$) and test × time interaction ($P < 0.001$) were found for glucose R_aO calculated with the 2-CM approach using a constant pV of 130 mL/kg. Data are mean ± SEM. ***$P < 0.001$, **$P < 0.01$ ISO vs. SUC

4.1.6.4 Endogenous Glucose Production

EGP was comparable in all subjects prior to ingestion of ISO or SUC (~8 μmol-/kg/min, $P = 0.790$). After ISO bolus as depicted in Fig. 4.14, EGP decreased to a nadir close to zero at approximately 120 min ($P < 0.01$ vs. preload value), and thereafter returned to the preload rate between 180 and 240 min. In contrast, EGP increased initially to a peak following ~60 min of SUC intake ($P < 0.01$ vs. preload value), soon decreased to near zero between 90 and 105 min, and then rose again to near preload value by ~180 min. Accordingly, mean EGP during the 4-h postprandial period was significantly lower after ISO load than after the SUC load ($P < 0.001$). Total amount of EGP was nearly doubled after intake of SUC compared with ISO ($P < 0.001$).

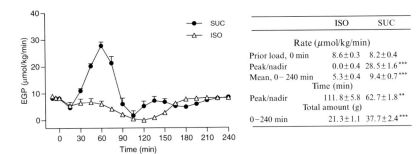

Fig. 4.14 EGP after ingestion of [1-^{13}C]ISO or [1-^{13}C]SUC in 11 T2DM subjects. A significant test effect ($P < 0.001$) and test × time interaction ($P < 0.001$) were found for EGP calculated with the 2-CM approach using a constant pV of 130 mL/kg. Data are mean ± SEM. ***$P < 0.001$, **$P < 0.01$ ISO vs. SUC

4.1.6.5   Total Glucose Rates of Disappearance

The time course of plasma glucose R_dT after ingestion of both disaccharides is shown in Fig. 4.15, which followed a similar pattern to that of glucose R_aT. Prior to ingestion, glucose R_dT was similar on both ISO and SUC study days ($\sim$9 μmol/kg/min, $P = 0.790$). Glucose R_dT increased after the SUC load ($P < 0.001$), peaking at $\sim$60 min, and decreased thereafter to near preload values by 180 and 240 min. Following ISO bolus, glucose R_dT also increased significantly ($P < 0.001$), reaching a maximum value at a later time point ($\sim$75 min, $P = 0.033$), and returned to the preload values by the end of the postprandial period.

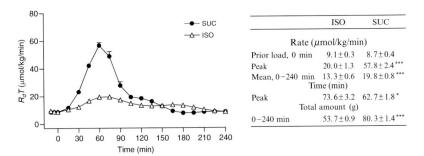

Fig. 4.15 Glucose R_dT after ingestion of [1-^{13}C]ISO or [1-^{13}C]SUC in 11 T2DM subjects. A significant test effect ($P < 0.001$) and test × time interaction ($P < 0.001$) were found for glucose R_dT calculated with the 2-CM approach using a constant pV of 130 mL/kg. Data are mean ± SEM. ***$P < 0.001$, *$P < 0.05$ ISO vs. SUC

Overall, the rates were lower after ISO ingestion compared with those following SUC ingestion ($P < 0.001$), with a $\sim$65 % lower peak after the ISO bolus ($P < 0.001$). As a result, mean glucose R_dT was significantly lower after ISO versus SUC intake ($P < 0.001$). During the 4-h postprandial period, the magnitude of oral and endogenous glucose entering the systemic circulation was reduced after ISO versus SUC bolus. Accordingly, approximately 27 g lesser quantity of glucose needed to be disposed from the systemic circulation after the ISO load ($P < 0.001$).

4.1.6.6 Splanchnic Glucose Uptake

Subjects consumed an average amount of 90.1 ± 4.6 g of ISO or SUC. Since ISO and SUC consist of 50 % glucose and 50 % fructose, oral glucose load is equivalent to half of the amount of ISO or SUC ingested. This corresponded to a total load of 45.0 ± 2.3 g of oral glucose. Of these, approximately 70 % or 90 % reached the systemic circulation respectively after ISO or SUC intake. Accordingly, SGU was enhanced after ISO compared with SUC administration (Table 4.2, $P = 0.003$).

Table 4.2 SGU after ingestion of [1-^{13}C]ISO or [1-^{13}C]SUC in 11 T2DM subjects

	ISO	SUC
SGU (g)	14.5 ± 2.5	4.3 ± 2.0**
SGU (%)	30.6 ± 4.3	8.3 ± 3.7**

Data are mean $\pm$ SEM
**$P < 0.01$ ISO vs. SUC

4.1.6.7 One-Compartment Versus Two-Compartment Models

Figures 4.16, 4.17, 4.18, and 4.19 show the changing patterns of glucose R_aT, glucose R_aO, EGP, and glucose R_dT estimated by the 1-CM and 2-CM approaches using a constant pV of 130 mL/kg, respectively. Both models provided a similar changing pattern for all parameters. However, estimates using 1-CM were overall lower than those calculated with the 2-CM, being more pronounced following the SUC load.

As depicted in Fig. 4.16, glucose R_aT reached a $\sim$15–20 % lower peak following administration of both disaccharides when calculated with the 1-CM ($P < 0.001$ vs. 2-CM). With the 2-CM method, both mean rate and total amount of glucose appeared in the systemic circulation during the 4-h postprandial period were significantly higher than those calculated using the 1-CM, either after ingestion of ISO or SUC ($P < 0.05$).

Similar to the glucose R_aT profile, glucose R_aO attained an approximately 10–15 % lower maximum value after ingestion of the disaccharides when estimated using the 1-CM approach compared with the 2-CM method ($P < 0.001$, Fig. 4.17). Moreover, during the 4-h postprandial period, mean rate of oral glucose appearance as well as the cumulative amount of oral glucose originating from each disaccharide were significantly underestimated using the 1-CM ($P < 0.001$ vs. 2-CM).

Calculation of EGP using the 1-CM approach, again, resulted in a significant underestimation of the peak value by approximately 28 % following administration

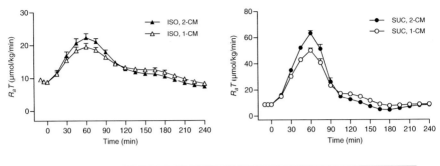

	ISO		SUC	
	1-CM	2-CM	1-CM	2-CM
Peak rate (μmol/kg/min)	19.6 ± 1.1	$22.5\pm1.2^{***}$	50.6 ± 2.0	$63.8\pm2.2^{***}$
Mean rate, $0-240$ min (μmol/kg/min)	13.1 ± 0.6	$13.2\pm0.6^{**}$	19.0 ± 0.7	$19.8\pm0.8^{***}$
Total amount, $0-240$ min (g)	53.1 ± 0.9	$53.5\pm0.8^{*}$	76.8 ± 1.2	$80.1\pm1.3^{***}$

Fig. 4.16 Comparison of glucose R_aT calculated using 1-CM and 2-CM approaches after ingestion of [1-^{13}C]ISO or [1-^{13}C]SUC in 11 T2DM subjects. For both models, glucose R_aT was calculated using a constant pV of 130 mL/kg. Data are mean $\pm$ SEM. $^{***}P < 0.001$, $^{**}P < 0.01$, and $^{*}P < 0.05$ 1-CM vs. 2-CM

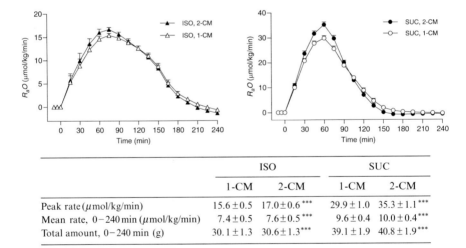

	ISO		SUC	
	1-CM	2-CM	1-CM	2-CM
Peak rate (μmol/kg/min)	15.6 ± 0.5	$17.0\pm0.6^{***}$	29.9 ± 1.0	$35.3\pm1.1^{***}$
Mean rate, $0-240$ min (μmol/kg/min)	7.4 ± 0.5	$7.6\pm0.5^{***}$	9.6 ± 0.4	$10.0\pm0.4^{***}$
Total amount, $0-240$ min (g)	30.1 ± 1.3	$30.6\pm1.3^{***}$	39.1 ± 1.9	$40.8\pm1.9^{***}$

Fig. 4.17 Comparison of glucose R_aO calculated using 1-CM and 2-CM approaches after ingestion of [1-^{13}C]ISO or [1-^{13}C]SUC in 11 T2DM subjects. For both models, glucose R_aO was calculated using a constant pV of 130 mL/kg. Data are mean $\pm$ SEM. $^{***}P < 0.001$ 1-CM vs. 2-CM

of SUC ($P < 0.001$ vs. 2-CM, Fig. 4.18). The 1-CM indicated that a smaller quantity of endogenous glucose was released into the systemic circulation during the 4-h postprandial period than the 2-CM approach ($P < 0.001$). Mean EGP

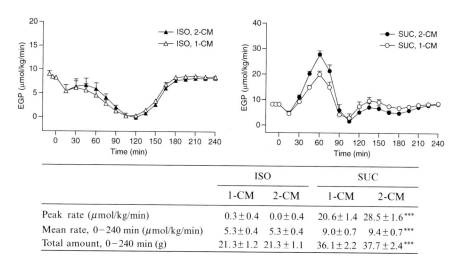

	ISO		SUC	
	1-CM	2-CM	1-CM	2-CM
Peak rate (μmol/kg/min)	0.3±0.4	0.0±0.4	20.6±1.4	28.5±1.6***
Mean rate, 0–240 min (μmol/kg/min)	5.3±0.4	5.3±0.4	9.0±0.7	9.4±0.7***
Total amount, 0–240 min (g)	21.3±1.2	21.3±1.1	36.1±2.2	37.7±2.4***

Fig. 4.18 Comparison of EGP calculated using 1-CM and 2-CM approaches after ingestion of [1-^{13}C]ISO or [1-^{13}C]SUC in 11 T2DM subjects. For both models, EGP was calculated using a constant pV of 130 mL/kg. Data are mean ± SEM. ***$P < 0.001$ 1-CM vs. 2-CM

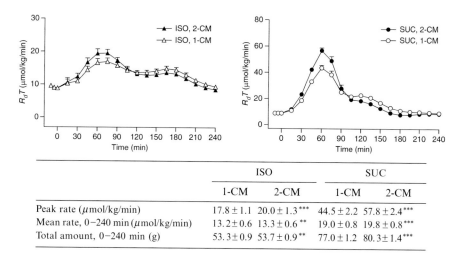

	ISO		SUC	
	1-CM	2-CM	1-CM	2-CM
Peak rate (μmol/kg/min)	17.8±1.1	20.0±1.3***	44.5±2.2	57.8±2.4***
Mean rate, 0–240 min (μmol/kg/min)	13.2±0.6	13.3±0.6**	19.0±0.8	19.8±0.8***
Total amount, 0–240 min (g)	53.3±0.9	53.7±0.9**	77.0±1.2	80.3±1.4***

Fig. 4.19 Comparison of glucose R_dT calculated using 1-CM and 2-CM approaches after ingestion of [1-^{13}C]ISO or [1-^{13}C]SUC in 11 T2DM subjects. For both models, glucose R_dT was calculated using a constant pV of 130 mL/kg. Data are mean ± SEM. ***$P < 0.001$, **$P < 0.01$ 1-CM vs. 2-CM

value followed the same pattern as the total amount ($P < 0.001$). In contrast, EGP estimations by the 1-CM and 2-CM were comparable after ingestion of ISO.

Consistent with the results of glucose R_aT, approximately 10 % and 20 % higher peak values of glucose R_dT were observed respectively following ISO and SUC loads when calculated using the 2-CM versus the 1-CM ($P < 0.001$, Fig. 4.19).

With the 2-CM, both mean rate and total amount of glucose leaving the systemic circulation during the 4-h postprandial period were significantly higher than those calculated using the 1-CM ($P < 0.01$).

Consistent with these results, a previous study using triple-tracer technique also indicated that the 1-CM approach underestimates postprandial glucose fluxes (Basu et al. 2003). Thus, the 2-CM is more accurate for calculation of glucose kinetics than the 1-CM under non-steady-state condition.

4.1.6.8 Fractional [6,6-²H₂]glucose Rates of Disappearance

Infusion of [6,6-^{2}H$_2$]glucose tracer allows calculation of the fractional disappearance rates of [6,6-^{2}H$_2$]glucose, which can be estimated from the mass balance equation of the tracer itself. When calculated with the 1-CM approach using a constant pV of 130 mL/kg, the fractional clearance rates of [6,6-^{2}H$_2$]glucose, denoted by $k_{01,2\mathrm{H}}$ in Fig. 4.20 (left panel), decreased immediately following ingestion of ISO and reached a nadir between 90 and 120 min. By the end of the period between 180 and 240 min, the previous load rate was reached again. Since [6,6-^{2}H$_2$]glucose tracer was used to trace the systemic glucose turnover, it was assumed that $k_{01,2\mathrm{H}}$ should follow the profile pattern of systemic glucose R_d as defined by the ratio of plasma glucose-to-insulin concentration multiplied by a constant factor (Best et al. 1981). The result showed that the systemic glucose R_d for the reference of $k_{01,2\mathrm{H}}$ (ref) had a different changing pattern following the ISO load. In contrast to the estimates using the 1-CM, the $k_{01,2\mathrm{H}}$ reference began to accelerate soon after ingestion of ISO and reached a plateau before returning to the preload rate by the end of the 4-h postprandial period.

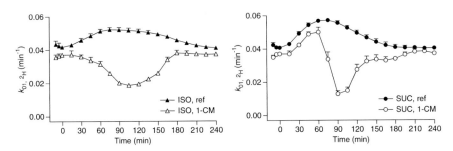

Fig. 4.20 Fractional $k_{01,2\mathrm{H}}$ after ingestion of [1-^{13}C]ISO or [1-^{13}C]SUC in 11 T2DM subjects. $k_{01,2\mathrm{H}}$ was calculated using the 1-CM approach with a constant pV of 130 mL/kg and using the assumption that $k_{01,2\mathrm{H}}$ follows the profile pattern of systemic glucose R_d, denoted as reference (ref). Data are mean ± SEM

Figure 4.20 (right panel) shows that during the first hour following ingestion of SUC, the fractional $k_{01,2\mathrm{H}}$ calculated with the 1-CM and fixed pV rose steadily, subsequently decreased to a minimum value by 90 min, and increased again continuously to reach the prior ingestion rate by 240 min. In contrast, $k_{01,2\mathrm{H}}$

reference increased gradually to reach a peak by 75 min, and returned slowly to the preload rate by the end of the 4-h period.

Thus, the above results indicate that calculation with the 1-CM approach and a constant pV was inaccurate to describe the postprandial tracer and tracee kinetics. By using the reference pattern of $k_{01,2H}$, values of pV at each time point can be determined. As shown in Fig. 4.21, the calculated pV values varied over time, ranging from 50 to 130 mL/kg.

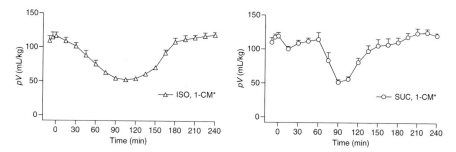

Fig. 4.21 Time-varying pV after ingestion of $[1-^{13}C]ISO$ or $[1-^{13}C]SUC$ in 11 T2DM subjects. pV was calculated using the mass balance equation of $[6,6-^2H_2]$glucose and needs to be used with the 1-CM approach to follow the same pattern of $k_{01,2H}$ as the reference. 1-CM* = 1-CM with a time-varying pV. Data are mean $\pm$ SEM

Figures 4.22 and 4.23 show the time course of plasma glucose R_aT and glucose R_dT calculated with the 1-CM and the variable pV method (designated as 1-CM*)

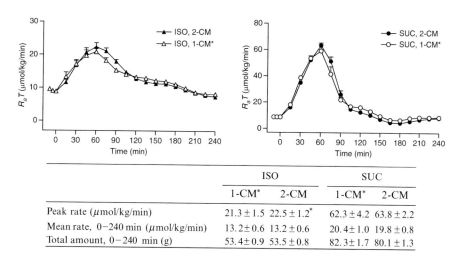

	ISO		SUC	
	1-CM*	2-CM	1-CM*	2-CM
Peak rate (μmol/kg/min)	21.3±1.5	22.5±1.2*	62.3±4.2	63.8±2.2
Mean rate, 0–240 min (μmol/kg/min)	13.2±0.6	13.2±0.6	20.4±1.0	19.8±0.8
Total amount, 0–240 min (g)	53.4±0.9	53.5±0.8	82.3±1.7	80.1±1.3

Fig. 4.22 Comparison of glucose R_aT calculated with the 1-CM using a time-varying pV and the 2-CM using a fixed pV after ingestion of $[1-^{13}C]ISO$ or $[1-^{13}C]SUC$ in 11 T2DM subjects. 1-CM* = 1-CM with a time-varying pV. Data are mean $\pm$ SEM. *$P < 0.05$ 1-CM vs. 2-CM

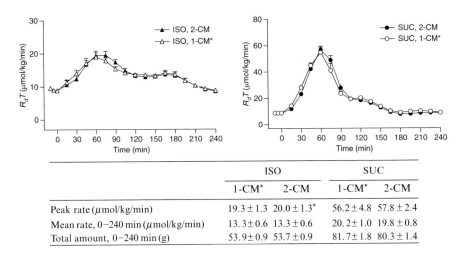

	ISO		SUC	
	1-CM*	2-CM	1-CM*	2-CM
Peak rate (μmol/kg/min)	19.3 ± 1.3	$20.0 \pm 1.3^*$	56.2 ± 4.8	57.8 ± 2.4
Mean rate, 0–240 min (μmol/kg/min)	13.3 ± 0.6	13.3 ± 0.6	20.2 ± 1.0	19.8 ± 0.8
Total amount, 0–240 min (g)	53.9 ± 0.9	53.7 ± 0.9	81.7 ± 1.8	80.3 ± 1.4

Fig. 4.23 Comparison of glucose $R_d T$ calculated with the 1-CM using a time-varying pV and the 2-CM using a fixed pV after ingestion of [1-^{13}C]ISO or [1-^{13}C]SUC in 11 T2DM subjects. 1-CM* = 1-CM with a time-varying pV. Data are mean $\pm$ SEM. $^*P < 0.05$ 1-CM vs. 2-CM

in comparison with the 2-CM and a constant pV approach. The changing patterns of glucose $R_a T$ and glucose $R_d T$ were nearly identical throughout the 4-h postprandial period when estimated using the both methods. Glucose $R_a T$ and glucose $R_d T$ began to rise from a similar rate prior to ingestion of ISO or SUC ($\sim$9 μmol/kg/min), subsequently reached a peak that was only marginally higher in the 2-CM at $\sim$60 min after the ISO load but was similar between the 1-CM and 2-CM after the SUC bolus, and returned continuously over the time to near preload rates by the end of the experiment. Mean glucose $R_a T$ and glucose $R_d T$ were similar during the 4-h postprandial period after ISO or SUC intake. Consequently, cumulative amounts of glucose entering and leaving the circulation after each disaccharide load were not different between the two approaches.

Based on the results above, the 1-CM approach with the fixed pV is inaccurate to describe the postprandial glucose turnover following oral administration of ISO or SUC. The variable pV approach for the calculation of systemic postprandial glucose kinetics under non-steady-state condition, as is evident from Figs. 4.22 and 4.23, is superior for this application. The underestimation of glucose turnover calculated with the 1-CM and fixed pV can be minimized by varying the pV values, yielding results that were virtually indistinguishable as compared with the 2-CM approach.

4.1.7   Insulin Sensitivity Index

Net S_I was higher after ingestion of ISO than after ingestion of SUC (Table 4.3, $P = 0.011$). A $\pm20\%$ deviation of glucose effectiveness produced only a small change in S_I (range: 12.6–13.9 and 4.6–4.7 $\times 10^{-4}$ dL/kg·min per pmol/L respectively for ISO and SUC).

Table 4.3 S_I after ingestion of [1-^{13}C]ISO or [1-^{13}C]SUC in 11 T2DM subjects

	ISO	SUC
S_I (10^{-4} dL/kg·min per pmol/L)	13.3 ± 3.2	4.7 ± 0.9*

Data are mean $\pm$ SEM
*$P < 0.05$ ISO vs. SUC

4.2 Isomaltulose-Protein Study

4.2.1 Subjects' Characteristics

The characteristics of the subjects are summarized in Table 4.4. Thirty T2DM subjects participated in this study. They were men and women aged between 45 and 75 years. Of these, 60 % were men. They had a mean diabetes duration of approximately 5 years, were overweight, but had good glycemic control with HbA$_{1c}$ less than 53 mmol/mol (7 %). Eleven of the subjects were treated with diet alone, and the rest were under metformin regimen or in combination with diet. No other anti-diabetic medication was used. Metformin administration was stopped 3 days prior to the tests. None of the subjects had a history of unstable or untreated proliferative retinopathy, clinically significant nephropathy, neuropathy, hepatic disease, heart failure, uncontrolled hypertension, systemic treatment with corticosteroids, and insulin treatment. Subjects were insulin resistant and had a β-cell function of approximately 70 %.

Table 4.4 Subjects' characteristics

Number of subjects	30
Sex (female/male)	12/18
Age (years)	62.9 ± 1.3
Body weight (kg)	85.6 ± 3.1
Height (m)	1.7 ± 0.0
BMI (kg/m^2)	29.0 ± 0.7
HbA$_{1c}$ (mmol/mol)	48.2 ± 1.4
HbA$_{1c}$ (%)	6.6 ± 0.1
Fasting plasma glucose level (mmol/L)[a, b]	7.2 ± 0.3
Fasting plasma insulin level (pmol/L)[a, c]	82.5 ± 11.5
HOMA-IR[d]	4.0 ± 0.6
HOMA-β (%)	71.1 ± 10.7
Diabetes duration (years)	5.2 ± 2.6
Treatment (diet/metformin)	11/19
Metformin (female/male)	8/11

Data are mean $\pm$ SEM
[a]Values prior to ingestion of ISO+WS and ISO+C
[b]Measurement was in mg/dL and converted to mmol/L by dividing by 18
[c]Measurement was in mU/L and converted to pmol/L by multiplying by 6.945
[d]A cut-off value of >2.60 is considered insulin resistance (Ascaso et al. 2003)

4.2.2 Plasma Glucose Concentrations

Fasting plasma glucose concentrations did not differ between the study days on which only ISO, ISO+WS, or ISO+C was ingested ($P = 0.648$). After consumption of all the three drinks, plasma glucose concentrations increased above the basal levels (all $P < 0.001$), reached peak values at ~75 min, and returned to baseline by ~240 min. At every time point, postprandial plasma glucose levels did not differ significantly between all the ingestions ($P = 0.411$, Fig. 4.24). Consequently, mean plasma glucose concentrations and plasma glucose iAUCs were not significantly different following administration of the different drinks ($P = 0.704$ and $P = 0.239$, respectively).

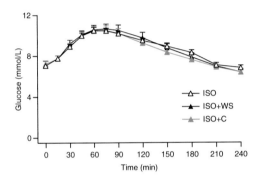

	ISO	ISO+WS	ISO+C	Overall P
Basal level (mmol/L)	7.1 ± 0.2	7.2 ± 0.3	7.2 ± 0.3	0.648
Peak level ((mmol/L)	11.2 ± 0.3	11.1 ± 0.4	10.9 ± 0.4	0.656
Peak time (min)	77.5 ± 4.0	78.5 ± 4.4	76.0 ± 4.0	0.639
Mean level, 0 –240 min (mmol/L)	8.8 ± 0.2	8.8 ± 0.4	8.6 ± 0.4	0.704
iAUC, 0–240 min (mmol/L·min)	428.2 ± 28.1	431.1 ± 41.9	385.0 ± 33.7	0.239

Fig. 4.24 Plasma glucose concentrations after ingestion of ISO alone, ISO+WS, and ISO+C in 30 T2DM subjects. No significant test effect ($P = 0.704$) or test × time interaction ($P = 0.411$) was found. Data are mean ± SEM

4.2.3 Plasma Insulin Concentrations

Fasting plasma insulin concentrations did not differ between the days of ingestion of ISO alone, ISO+WS, and ISO+C ($P = 0.525$). Plasma insulin concentrations increased from baseline to peak levels after ~90 min consumption of the different drinks (all $P < 0.001$) and returned to baseline by ~240 min. Significant differences in postprandial insulin levels were observed following ingestion of the drinks ($P < 0.001$, Fig. 4.25). Peak insulin concentration was higher after administration of ISO+WS or ISO+C than after administration of ISO only (both $P < 0.001$), and was elevated following intake of ISO+WS compared to ISO+C ($P < 0.001$). As a result, plasma insulin responses were respectively ~270% and ~190% greater after ingestion of ISO+WS and ISO+C than after ingestion of ISO alone (both $P < 0.001$), and was ~30% enhanced following consumption of ISO+WS versus ISO+C ($P = 0.006$). Mean plasma insulin concentrations followed the same pattern as plasma responses ($P < 0.001$).

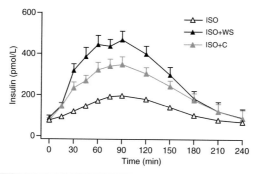

	ISO	ISO+WS	ISO+C	Overall P
Basal level (pmol/L)	77.1 ± 2.9	88.6 ± 11.2	76.5 ± 11.8	0.525
Peak level (pmol/L)	212.4 ± 10.2	571.9 ± 47.0***	445.4 ± 38.6 *** †††	<0.001
Peak time (min)	94.0 ± 4.1	94.0 ± 7.3	95.5 ± 10.0	0.515
Mean level, 0–240 min (pmol/L)	131.1 ± 5.2	282.4 ± 20.1***	222.5 ± 15.6 *** †††	<0.001
iAUC, 0–240 min (pmol/L·min)	13819.3 ± 1093.2	51346.3 ± 3960.9 ***	39957.6 ± 3013.7*** ††	<0.001

Fig. 4.25 Plasma insulin concentrations after ingestion of ISO alone, ISO+WS, and ISO+C in 30 T2DM subjects. A significant test effect ($P < 0.001$) and test × time interaction ($P < 0.001$) were found. Data are mean ± SEM. P values for multiple comparisons were adjusted using Bonferroni correction. ***$P < 0.001$ ISO vs. ISO+WS and ISO vs. ISO+C. †††$P < 0.001$, ††$P < 0.01$ ISO+WS vs. ISO+C

4.2.4 Insulin Sensitivity Index

Net S_I was lower after ingestion of ISO+WS and ISO+C than after administration of ISO alone ($P < 0.001$ and $P = 0.015$, respectively). Moreover, S_I was lower following intake of ISO+WS compared to that following intake of ISO+C ($P = 0.009$, Table 4.5).

Table 4.5 S_I after ingestion of ISO alone, ISO+WS, and ISO+C in 30 T2DM subjects

	ISO	ISO+WS	ISO+C	Overall P
S_I (10^{-5} dL/kg·min per pmol/L)	16.3 ± 2.0	$5.9 \pm 0.8^{***}$	$10.4 \pm 1.8^{*\dagger\dagger}$	<0.001

Data are mean $\pm$ SEM

P values for multiple comparisons were adjusted using Bonferroni correction

$^{***}P < 0.001$, $^{*}P < 0.05$ ISO vs. ISO+WS and ISO vs. ISO+C. $^{\dagger\dagger}P < 0.01$ ISO+WS vs. ISO+C

4.2.5 Plasma Amino Acid Concentrations

Fasting plasma concentrations of total amino acids (TAA) did not differ on the study days when ISO+WS and ISO+C were ingested ($P = 0.651$). As expected, plasma TAA concentrations increased from baseline following consumption of both drinks ($P < 0.001$) but were higher after administration of ISO+WS compared to those of ISO+C ($P < 0.001$, Fig. 4.26). Similarly, fasting plasma concentrations of essential amino acids (EAA) and branched-chain amino acids (BCAA) did not differ on the ISO+WS and ISO+C study days ($P = 0.221$ and $P = 0.141$, respectively). Plasma

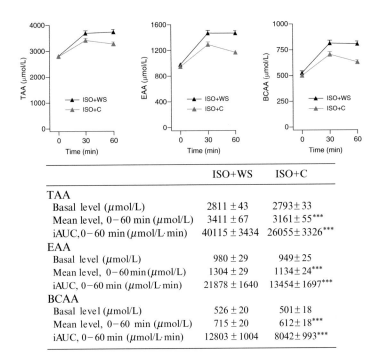

	ISO+WS	ISO+C
TAA		
Basal level (μmol/L)	2811 ± 43	2793 ± 33
Mean level, 0–60 min (μmol/L)	3411 ± 67	$3161 \pm 55^{***}$
iAUC, 0–60 min (μmol/L·min)	40115 ± 3434	$26055 \pm 3326^{***}$
EAA		
Basal level (μmol/L)	980 ± 29	949 ± 25
Mean level, 0–60 min (μmol/L)	1304 ± 29	$1134 \pm 24^{***}$
iAUC, 0–60 min (μmol/L·min)	21878 ± 1640	$13454 \pm 1697^{***}$
BCAA		
Basal level (μmol/L)	526 ± 20	501 ± 18
Mean level, 0–60 min (μmol/L)	715 ± 20	$612 \pm 18^{***}$
iAUC, 0–60 min (μmol/L·min)	12803 ± 1004	$8042 \pm 993^{***}$

Fig. 4.26 Plasma TAA, EAA, and BCAA concentrations after ingestion of ISO+WS and ISO+C in 30 T2DM subjects. A significant test effect ($P < 0.001$) and test × time interaction ($P < 0.001$) were found respectively for plasma TAA, EAA, and BCAA concentrations. Data are mean $\pm$ SEM. $^{***}P < 0.001$ ISO+WS vs. ISO+C

EAA and BCAA levels increased from baseline ($P < 0.001$) and showed a pattern similar to plasma TAA levels (Fig. 4.26). Moreover, both plasma EAA and BCAA levels were higher after administration of ISO+WS than after administration of ISO+C ($P < 0.001$). Plasma TAA concentrations correlated positively with plasma insulin levels in the 60-min postprandial period following ingestion of ISO+WS and ISO+C ($r_s = 0.690$ and $r_s = 0.684$; $P < 0.001$, respectively).

Plasma responses of TAA, EAA, and BCAA were respectively 54 %, 63 %, and 59 % higher after administration of ISO+WS than after administration of ISO+C (all $P < 0.001$). Plasma EAA and BCAA responses accounted for ~55 % and ~30 % of the increases in plasma TAA responses following consumption of both drinks, respectively. Mean plasma concentrations of TAA, EAA, and BCAA consistently followed the same pattern as their plasma responses (all $P < 0.001$).

An overview of individual plasma amino acid concentrations at 0, 30, and 60 min is shown in Table 4.6. After ingestion of ISO+WS, most plasma amino

Table 4.6 Plasma amino acid concentrations and iAUCs after ingestion of ISO+WS and ISO+C in 30 T2DM subjects

	ISO+WS (μmol/L)			ISO+C (μmol/L)			ISO+WS vs. ISO+C	
	Baseline	30 min	60 min	Baseline	30 min	60 min	iAUC (μmol/L·min)	
Ala	473 ± 20	621 ± 32	676 ± 29	483 ± 20	588 ± 27	631 ± 26	7501 ± 781	5526 ± 761**
Arg	79 ± 3	116 ± 5	108 ± 5	76 ± 3	96 ± 5	86 ± 4	1528 ± 149	761 ± 130***
Asn	43 ± 1	75 ± 4	75 ± 4	44 ± 1	61 ± 2	56 ± 3	1430 ± 133	698 ± 90***
Asp	10 ± 1	13 ± 1	12 ± 1	10 ± 1	11 ± 1	10 ± 1	142 ± 23	55 ± 9**
Cit	35 ± 2	34 ± 2	35 ± 2	33 ± 2	34 ± 2	34 ± 3	32 ± 10	68 ± 16
Gln	457 ± 23	500 ± 24	509 ± 23	463 ± 21	495 ± 24	483 ± 25	2187 ± 315	1451 ± 250
Glu	160 ± 19	178 ± 16	175 ± 17	157 ± 14	171 ± 16	171 ± 16	1164 ± 266	1004 ± 210
Gly	208 ± 13	231 ± 15	228 ± 18	211 ± 14	221 ± 15	211 ± 16	1030 ± 221	423 ± 105***
His	77 ± 2	87 ± 2	91 ± 2	77 ± 2	88 ± 2	88 ± 2	543 ± 47	491 ± 59
Ile	88 ± 4	166 ± 7	168 ± 7	86 ± 4	131 ± 6	114 ± 5	3521 ± 263	1795 ± 219***
Leu	157 ± 6	274 ± 12	266 ± 9	151 ± 6	231 ± 10	195 ± 7	5127 ± 424	3048 ± 400***
Lys	177 ± 7	270 ± 10	264 ± 9	173 ± 5	235 ± 9	212 ± 6	4112 ± 291	2456 ± 333***
Met	27 ± 2	38 ± 2	36 ± 2	26 ± 1	40 ± 2	35 ± 1	455 ± 38	567 ± 81
Phe	64 ± 3	84 ± 4	81 ± 4	64 ± 3	81 ± 4	73 ± 4	846 ± 73	654 ± 104
Ser	108 ± 4	151 ± 6	147 ± 6	108 ± 4	140 ± 6	129 ± 5	1880 ± 194	1280 ± 156**
Tau	106 ± 9	115 ± 11	113 ± 11	112 ± 9	124 ± 11	119 ± 10	690 ± 230	744 ± 192
Thr	126 ± 5	187 ± 8	195 ± 8	125 ± 5	158 ± 6	152 ± 5	2846 ± 239	1422 ± 165***
Trp	59 ± 2	76 ± 2	80 ± 3	60 ± 2	68 ± 2	64 ± 2	808 ± 72	354 ± 59***
Tyr	74 ± 4	99 ± 5	99 ± 5	71 ± 3	100 ± 5	90 ± 5	1123 ± 121	1135 ± 193
Val	280 ± 10	372 ± 11	374 ± 10	265 ± 9	343 ± 12	321 ± 9	4154 ± 343	3198 ± 377*

Data are mean ± SEM

P values for multiple comparisons were adjusted using Bonferroni correction

***$P < 0.001$, **$P < 0.01$, and *$P < 0.05$ ISO+WS vs. ISO+C

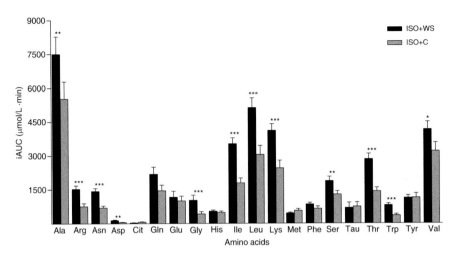

Fig. 4.27 Plasma amino acid iAUCs after ingestion of ISO+WS and ISO+C in 30 T2DM subjects. Data are mean ± SEM. *P* values for multiple comparisons were adjusted using Bonferroni correction. ****P* < 0.001, ***P* < 0.01, and **P* < 0.05 ISO+WS vs. ISO+C

acid concentrations increased significantly from the baseline level (*P* < 0.05), except plasma concentrations of citrulline, glutamic acid, and taurine. The highest increment was observed for plasma alanine, followed by plasma EAA and BCAA (i.e., plasma leucine, valine, lysine, isoleucine, and threonine). Similarly, all plasma amino acid levels also increased significantly above the baseline following ingestion of ISO+C (*P* < 0.05), except plasma concentrations of citrulline, glutamic acid, taurine, and glycine. Again, the highest increment was observed for plasma alanine, followed by valine, leucine, lysine, isoleucine, and threonine. Compared to ingestion of ISO+C, plasma responses of all these amino acids were significantly higher following ingestion of ISO+WS (*P* < 0.05, Fig. 4.27); approximately 30–40 % for plasma alanine, leucine, valine, and lysine, and approximately 95–100 % for plasma isoleucine and threonine.

Chapter 5
Discussion

5.1 Methodology

Plasma glucose concentrations are determined by the balance between rates of glucose entering and leaving the systemic circulation. Glucose reaching the circulation after ingestion of carbohydrate is derived from exogenous (meal) and endogenous (liver plus kidney) sources. The ISO/SUC-Clamp Study was undertaken to comprehensively characterize postprandial glucose changes after a bolus ingestion of ISO compared with SUC in T2DM patients. Using a combination of euglycemic-hyperinsulinemic clamp and labeled oral ISO or SUC load, glucose absorption and turnover of oral and endogenous glucose were assessed. This approach has several advantages over the other methods; three of them are mentioned as follows. It allows a straight discrimination between slowly and rapidly absorbed carbohydrates by directly determining the duration of glucose absorption from the experiment. It enables simultaneous quantification of rates of systemic glucose appearance and disappearance accurately as well as differentiation between oral and endogenous glucose release into the bloodstream. Lastly, extensive fluctuations in blood glucose levels from digestion of simple carbohydrates and recycling of C3-fragments in the liver are stabilized.

Prior to administration of ISO or SUC, infusions of insulin and $[6,6-^2H_2]$glucose tracer were started in parallel. Insulin was infused in a continuous fashion in order to decrease plasma glucose levels into euglycemic state (from $\sim$7.5 to 5 mmol/L) and maintain the levels at this state, while infusion of $[6,6-^2H_2]$glucose tracer enabled estimation of systemic glucose appearance. During insulin infusion, plasma glucose concentrations were maintained at the euglycemic range by concomitant infusion of glucose that was enriched with $[6,6-^2H_2]$glucose (also called hot-GINF method) to minimize changes in plasma tracer enrichments (Finegood et al. 1987). Equilibria for plasma insulin, glucose, and glucose tracer concentrations were achieved approximately 30 min before the oral load of ISO or SUC, as evidenced by the constant values of plasma $[6,6-^2H_2]$glucose ($\sim$2 %), plasma glucose ($\sim$5 mmol/L),

M. Ang, *Metabolic Response of Slowly Absorbed Carbohydrates in Type 2 Diabetes Mellitus*, SpringerBriefs in Systems Biology, DOI 10.1007/978-3-319-27898-8_5

and plasma insulin ($\sim$350 pmol/L) during the time period from -30 to 0 min. The time required for plasma insulin to reach the steady-state level in T2DM patients was considerably longer compared to other studies under similar constant insulin infusions in healthy or T1DM subjects (2.5 h vs. 1 h, respectively) (Meyer et al. 2005; Ang et al. 2014). The possible reason for this observation may be due to obesity or impaired insulin action in these T2DM patients. Most important, isotopic and substrate equilibriums were attained prior to administration of both disaccharides, which are prerequisites for measuring glucose kinetics accurately.

To allow measurement of dynamic metabolic responses discriminating both disaccharides, insulin was administered at a rate of 0.8 mU/kg/min, which resulted in approximate target insulin concentrations of 350 pmol/L. Using this infusion rate, basal C-peptide secretion was readily suppressed by more than half. Insulin infusion at a rate of 1.0 mU/kg/min for 2 h has been demonstrated to increase plasma insulin levels to $\sim$700 pmol/L, which inhibits EGP completely or almost completely in healthy and T2DM subjects (Rizza et al. 1981; Campbell et al. 1988). Accordingly, any higher rate than 1.0 mU/kg/min of insulin infusion will entirely inhibit endogenous C-peptide release as well as EGP even prior to administration of both disaccharides. This, in turn, would exacerbate dynamic metabolic assessment associated with the absence of changes in glucose kinetics subsequent to their intake. On the other hand, when using a lower insulin infusion, peripheral glucose uptake will not be highly stimulated in insulin-resistant T2DM subjects. However, even at a higher rate of insulin infusion, glucose disposal is still relatively low in T2DM as opposed to healthy probands (Campbell et al. 1988). Based on these considerations, opting for insulin infusion rate less than 1 mU/kg/min is favorable to allow the feasibility of estimating glucose kinetics following ISO or SUC load. Moreover, different degrees of impairment of both insulin action and/or endogenous insulin secretion in T2DM individuals were adjusted prior to intake of both disaccharides.

Subjects in the ISO/SUC-Clamp Study were mostly obese, with a mean BMI of $\sim$32 kg/m^2. In the experiments, the probands consumed 1 g/kg body weight of ^{13}C-enriched ISO or SUC. Their average body weight was 90 kg (range: 72–119 kg), which means that they ingested an average amount of 90 g of disaccharides. This corresponds to total daily sugar intake in most developed countries, ranging from 85 to 125 g or approximately 15–20 % of daily energy intake (Sugar Nutrition UK 2011). Therefore, the amounts of disaccharides ingested by the subjects represent daily sugar consumption.

After ISO or SUC intake, exogenous GINF was reduced subsequently from the preload rates in order to compensate for oral glucose absorption and to maintain euglycemia. Due to insulin resistance of the subjects, a decrease in exogenous GINF following ISO or SUC ingestion did not fully accommodate to the absorbed oral glucose; plasma glucose levels rose consequently. These results are consistent with protocols of similar condition in T2DM subjects (Ludvik et al. 1997; Bajaj et al. 2003). Labeling ISO and SUC, however, allowed postprandial oral glucose appearance to be accurately measured and distinguished from endogenous glucose release into the systemic circulation.

Postprandial glucose flux can be accurately calculated when tracer and tracee concentrations are constant, because the assumption of uniform mixing of tracer with tracee in the compartment applies and no recycling of tracer occurs once it has left the system (Steele et al. 1956; Vella and Rizza 2009). To anticipate, minimize, and stabilize changes in fluctuation pattern of both tracer and tracee concentrations, oral disaccharides load was therefore combined with hyperinsulinemic-euglycemic clamp in conjunction with the hot-GINF technique. Nevertheless, perturbation in steady-state plasma [6,6-^{2}H$_2$]glucose tracer occurred following administration of both disaccharides. To account for the possibility that glucose may not be homogeneously distributed in a single pool, postprandial glucose flux was therefore calculated under non-steady-state condition using 1-CM and 2-CM approaches in combination with Finegood's procedure. Compared to results computed with a 2-CM, the 1-CM underestimated glucose flux during non-steady state, but its accuracy can be increased by varying glucose distribution volume (pV) in the calculation. A smoothing procedure using OOPSEG program can further enhance the accuracy of the estimation in both models, since smoothing minimizes measurement inaccuracy by filtering noisy data and reducing deviation of the smoothed points. A previous study by Basu et al. (2003) also indicates that the 2-CM is precise in approximating glucose kinetics at disequilibrium and is superior to the 1-CM approach. Thus, mathematical modeling using the 2-CM combined with Finegood's procedure and OOPSEG smoothing as mentioned above lead to validated and reproducible results.

5.2 Glucose Absorption

Consistent with previous experimental results (Tsuji et al. 1986; Goda et al. 1988; Tonouchi et al. 2011), the ISO/SUC-Clamp Study showed that absorption of ISO was significantly prolonged compared with SUC. Absorption of each disaccharide was determined from oral glucose appearance in plasma, which was calculated by means of labeled [^{13}C]glucose data. Oral glucose appearance reflects ingested glucose that has been absorbed following the oral load of the ^{13}C-enriched disaccharides and is released into the circulation. A complete glucose absorption was indicated by the returned rates of oral glucose appearance to baseline values. This calculation method shows that glucose absorption from ISO was finished after ~211 min of ingestion, while absorption of SUC was more rapid and lasted ~161 min, indicating a delayed ISO absorption by approximately 50 min. Estimation of glucose absorption using GINF values also yielded similar results; glucose absorption was slowed down by ~40 min following ISO as opposed to SUC consumption.

Several factors may cause differences in the absorption rate of ISO and SUC. One important determinant is the discrepancy in their chemical structures. Compared with SUC, α-1,6 glycosidic bonds between glucose and fructose molecules in ISO are hydrolyzed more slowly (Dahlqvist et al. 1961, 1963; Goda and Hosoya 1983;

Tsuji et al. 1986; Goda et al. 1988), postponing contact of glucose with absorptive surface to a later time point. This is readily attributed to the lower activity of iso-maltase moiety of sucrase-isomaltase complex for cleaving ISO compared to that of sucrose (Dahlqvist et al. 1963; Goda and Hosoya 1983; Goda et al. 1988). A portion of ISO is also hydrolyzed by glucoamylase (Dahlqvist et al. 1961; Goda et al. 1988; Günther and Heymann 1998). In contrast, SUC is exclusively digested by sucrase. Both enzymes have a different kind of distribution throughout the small intestine; sucrase has its highest activity in the proximal jejunum and considerably lesser activity in the distal ileum, whereas glucoamylase activity increases regularly from pylorus to ileocecal valve (Triadou et al. 1983). Evidence suggests that a greater length of small intestine will be exposed to nutrients with increasing nutritional loads, which occurs due to saturation of digestive and absorptive capacities. As incoming loads exceed the absorptive capacity of the proximal segment of small intestine, more and more nutrients will be shifted beyond the segment, reaching the distal gut (Lin et al. 1989; Meyer et al. 1998b). Consequently, even if higher loads of ISO saturate jejunal isomaltase, undigested ISO spilling into ileum will be readily hydrolyzed by glucoamylase and thus contributes to even more prolonged absorption. By contrast, SUC can still be easily cleaved in the proximal small bowel, despite of a higher load because of increased sucrase distribution and activity in the upper than in the lower gut. Thus, differences in the duration of absorption between ISO and SUC will be even more pronounced as an increasing amount of ISO travels along the distal small gut.

Alterations in gastrointestinal motility as an adaptation to the digestion process may be another contributing factor to the slower absorption of ISO compared with SUC. Following digestion, glucose released from ISO will interact with glucose receptors in the small intestine. This, in turn, mediates a variety of gastrointestinal responses, which controls gastric distension, gastric motility, and gastrointestinal hormone secretion. This process, also known as feedback mechanism, serves for efficient digestion and absorption in the gastrointestinal tract (Ehrlein and Schemann 2005). Gastric emptying becomes inhibited once glucose resulting from digestion is available for absorption. Initial gastric emptying may be more rapid than the subsequent rate due to a delay in feedback inhibition by small intestine receptors. Both amounts of glucose loads and intestine's length exposed to glucose determine gastric emptying rate; the more glucose entering the distal gut, the greater the inhibition of stomach emptying (Horowitz et al. 1993; Schirra et al. 1996; Little et al. 2006). Feedback signals for delaying gastric emptying is controlled by enhanced release of GLP-1 hormone following ISO load, which contributes secondary to the slowed glucose absorption.

Low-digestible carbohydrates are incompletely or not absorbed in the small intestine, and are fermented by bacteria in the large intestine, causing gastroin-testinal discomforts including abdominal pain, flatulence, or diarrhea. Results in healthy subjects indicate that ISO is completely degraded and absorbed in the small intestine (Holub et al. 2010), thus no fermentation is expected in the colon (Kashimura et al. 1990; Tamura et al. 2004). Preliminary data with increasing ISO or SUC doses from 12.5 to 100 g demonstrated no significant differences

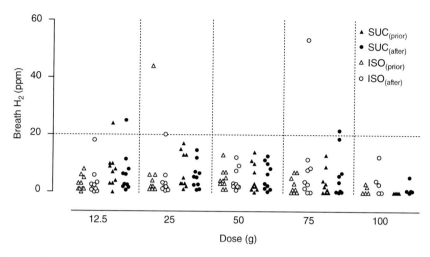

Fig. 5.1 Breath hydrogen concentrations prior to and after ingestion of 12.5, 25, 50, 75, and 100 g of ISO or SUC in T2DM subjects. Ten subjects ingested 12.5, 25, and 50 g of ISO or SUC in a crossover fashion. Similarly, 9 and 5 subjects consumed 75 and 100 g of ISO or SUC, respectively. After ingestion, breath H_2 levels were measured in a 30-min interval over a 3-h postprandial period and were expressed as mean values for each subject

in the mean breath hydrogen concentrations following ISO versus SUC intake within a 3-h period ($P = 0.500$–0.594). A rise of 10–20 parts per million (ppm) over baseline is considered to indicate malabsorption (Simrén and Stotzer 2006). However, mean breath H_2 contents mostly did not exceed 20 ppm (Fig. 5.1). Two patients had elevated fasting breath H_2 levels (>20 ppm), consequently their mean levels expectedly exceeded 20 ppm after respective intake of 12.5 g SUC or 25 g ISO. One patient had high levels of breath H_2 following intakes of 75 g ISO and SUC, presumably due to fructose intolerance. Overall, these results suggest that ISO is not fermented by bacteria in the colon, therefore it is fully absorbed in the small intestine. In addition, no remarkable gastrointestinal discomforts were recorded in the tolerance questionnaires after ingestion of increasing ISO doses (data not shown), which indicate that consumption of ISO is as tolerable as SUC.

5.3   Plasma Glucose Concentrations

Numerous studies have demonstrated lower postprandial plasma glucose and peak glucose concentrations following ISO consumption compared with SUC in healthy volunteers (Macdonald and Daniel 1983; Kawai et al. 1985, 1989; Liao et al. 2001; van Can et al. 2009; Holub et al. 2010; Maeda et al. 2013), impaired glucose-tolerant subjects (van Can et al. 2012), and T2DM patients (Kawai et al. 1989; Liao et al. 2001). In the majority of the studies, plasma glucose levels were significantly attenuated, which were most pronounced during the first hour following ingestion of ISO as opposed to SUC. Consistent with those findings, peak glucose level was reduced by approximately 25 % after the ISO load in the present study.

A variation exists in the magnitude of the reduction of postprandial glucose excursions between the present ISO/SUC-Clamp Study and the above indicated studies. The difference in glycemic response (expressed as total iAUC) after bolus administration of ISO versus SUC was greater in the present study compared to previous studies (Kawai et al. 1989; Liao et al. 2001; Holub et al. 2010) (~45 % vs. ~30–35 %, respectively). The explanation for this observation can be attributed to the different experimental design applied in the present study, including experimental condition and oral dose of ISO or SUC. The amounts of ISO and SUC ingested by the subjects were considerably higher in the present study compared to those studies (~90 vs. 50–75 g). The effect of ISO intake on postprandial glucose response seems to be dose-dependent compared to SUC. It appears that the magnitude differences become even extended with greater loads.

Several reasons can explain the overall lower plasma glucose concentrations observed following ISO versus SUC intake. First, following the delayed digestion of ISO, oral glucose was gradually absorbed into the portal vein, causing a smaller amount of glucose released into the bloodstream. Second, endogenous glucose output was suppressed to a higher extent, consequently reducing additional glucose that reached the systemic circulation after the ISO load. Third, during the first pass through the liver, more oral glucose from ISO was extracted by the splanchnic tissues, leaving less glucose to enter the blood circulation. Fourth, the combination of high dose and experimental condition may exert some synergistic effects on postprandial glucose metabolism, leading to pronounced differences in plasma glucose levels between ISO and SUC.

Ingestion of ISO+WS or ISO+C did not reduce plasma glucose responses despite the substantial increase in postprandial insulin levels compared to ingestion of ISO alone in T2DM subjects. The explanation for this observation is a reduced postprandial insulin action after intake of ISO+WS or ISO+C. In healthy probands, short-term elevations of plasma amino acids have been demonstrated to decrease glucose disposal, inhibit glucose transport or muscle insulin signaling, and enhance EGP by increasing glycogenolysis or gluconeogenesis (Krebs et al. 2002, 2003). It is therefore possible that protein components of ISO+WS and ISO+C decrease the effects of insulin in T2DM subjects. Reduced insulin action is counterbalanced by an increase in insulin secretion, thereby maintaining plasma glucose levels.

The effects of co-administration of various proteins and carbohydrates on plasma glucose concentrations are discussed controversially. Healthy probands often showed small, but significant absolute effects on plasma glucose AUCs when two carbohydrate and protein blends were compared (Manders et al. 2005, 2006), although the shape of plasma glucose profiles did not differ (van Loon et al. 2003). In T2DM patients, plasma glucose responses were either not significant (Gannon et al. 1992) or highly variable and the magnitude of glucose reduction was only moderate (Gannon et al. 1988; Manders et al. 2005, 2006). Combining carbohydrate with casein reduced glucose responses, but systemic glucose appearance and disappearance rates (Manders et al. 2005) as well as plasma glucose concentrations were not significantly different (Manders et al. 2006) compared to ingestion of carbohydrate only. A reduction of blood glucose response was further reported after whey intake for lunch compared to a ham meal without significant differences for breakfast with the same meals (Frid et al. 2005). Altogether, no definitive conclusion on the glucose-lowering effect of proteins can be drawn from those studies.

5.4   Plasma Hormone Concentrations

5.4.1   Insulin and C-Peptide

As indicated above, insulin infusion increased basal plasma insulin concentrations to physiological levels and decreased basal endogenous insulin secretion by approximately 60 % prior to ISO or SUC load. Subsequently after administration of each disaccharide, plasma insulin concentrations rose significantly from the preload levels, indicating an appropriate secretion of residual endogenous insulin and hence the rise in plasma insulin levels in response to the ingestion of both disaccharides. The different responses in plasma C-peptide and insulin levels following the oral load of both disaccharides confirm the previous findings that ISO consistently attenuated postprandial insulin concentrations and AUCs/iAUCs in healthy participants (Macdonald and Daniel 1983; Kawai et al. 1985, 1989; Liao et al. 2001; van Can et al. 2009; Holub et al. 2010; Maeda et al. 2013), in impaired glucose-tolerant subjects (van Can et al. 2012), and in T2DM patients compared to SUC (Kawai et al. 1989; Liao et al. 2001).

Ingestion of ISO+WS or ISO+C substantially increased postprandial insulin levels compared to ingestion of ISO alone in T2DM subjects. Numerous reports have confirmed a stable insulinotropic effect of proteins combined with glucose or maltodextrin in T2DM patients (Gannon et al. 1988, 1992; van Loon et al. 2003; Manders et al. 2005, 2006). Because the amounts of ingested ISO were identical (50 g) in the ISO-Protein Study, differences in insulin responses depend exclusively on the protein compositions displayed by the different amino acid profiles following ingestion of ISO+WS or ISO+C. Present results confirm a previous report on the stimulation of insulin by postprandial TAA levels (Calbet and MacLean 2002). Consequently, a greater increase in plasma TAA concentrations following consumption of ISO+WS promotes a faster release of insulin compared to intake of ISO+C.

5.4.2   Glucagon

Plasma glucagon levels increased significantly to lower values after ISO consumption compared to SUC. The explanation for this observation can be attributed to the higher GLP-1 secretion, because GLP-1 infusions have been demonstrated to reduce plasma glucagon levels in healthy subjects as well as in T1DM and T2DM patients (Gutniak et al. 1992). Additionally, insulin infusion may contribute to the decreased glucagon response, since physiological insulin levels declined plasma glucagon concentrations by $\sim$45 % prior to the ISO load.

Plasma glucagon responses increased significantly after ingestion of ISO+WS or ISO+C ($P < 0.001$, data not shown). The increases in glucagon levels after ingestion of both solutions were correlated to plasma TAA levels ($P < 0.05$), indicating that the presence of proteins enhances postprandial glucagon excursions. Consistent with plasma insulin responses, a greater glucagon response was observed following intake of ISO+WS, which could be attributed to the higher increase in plasma TAA levels compared to intake of ISO+C.

5.4.3   Incretins

Differences in the rate of glucose absorption trigger a different pattern of incretin release. Following ISO intake, plasma GLP-1 levels were significantly greater, whereas plasma GIP concentrations were lower as compared with those following SUC intake. These results are in agreement with an ISO study in healthy non-obese subjects (Maeda et al. 2013). Incretin secretion is related to ingestion of meal or meal-derived glucose, since GLP-1 and GIP release is not stimulated by intravenous glucose (Nauck et al. 1986; Herrmann et al. 1995) or sweeteners (Gregersen et al. 2004), and depends on rates of nutrient delivery into the small intestine (O'Donovan et al. 2004; Chaikomin et al. 2007).

Low GIP response can be primarily explained by the slower absorption of glucose originating from ISO. A good correlation between plasma GIP and profile pattern of oral exogenous glucose appearance was reported in a study in healthy subjects after consuming pasta or bread. Pasta was found to be digested slower than bread, eliciting a lower rise in GIP release (Eelderink et al. 2012). Varying the rate of GINF, which was administered intraduodenally to healthy and T2DM subjects, has also been shown to influence GIP release. When the initial rate was more rapid, GIP increased greater compared with a constant rate (O'Donovan et al. 2004). Moreover, reducing the frequency of duodenal flow events was reported to reduce glucose absorption and GIP secretion (Chaikomin et al. 2007). Together, these results suggest that glucose absorption rates are a vital determinant of GIP release in the small intestine. GIP is secreted from the enteroendocrine K-cells mostly located in the proximal small bowel (duodenum and jejunum). Presumably, only a small amount of glucose resulting from the slower digestion of ISO was absorbed in the upper part of the gut, causing a lower GIP response compared with that following the SUC bolus.

Animal studies have demonstrated that plasma GLP-1 levels increase substantially subsequent to the interaction of ISO or SUC with the ileal lumen, where the most abundant GLP-1-producing cells are distributed. Surprisingly, a higher GLP-1 release was observed following direct administration of SUC into the ileum of rats than that following the instillation of ISO. Because ISO or SUC per se did not trigger an increase in GLP-1 release in the GLP-1-producing enteroendocrine cell line (GLUTag cells), rapid digestion of SUC may have caused more glucose to accumulate in the ileal lumen, thereby producing a greater GLP-1 release (Hira et al. 2011). Studies using α-glucosidase inhibitors, such as acarbose, have shown a prolonged and enhanced GLP-1 response after SUC ingestion in healthy and T2DM subjects. Acarbose prevents rapid hydrolysis of SUC and moves the digestion of SUC to later time points, thereby shifting absorption of glucose from the jejunal to the ileal intestine (Qualmann et al. 1995; Seifarth et al. 1998). Due to reduced degradation of carbohydrates by acarbose, a portion of SUC remains undigested, leading to increased H_2 production in breath. As indicated above, glucose but not ISO or SUC itself induces a significant GLP-1 secretion in the GLP-1 producing cells, which emphasizes the importance of glucose presence in the gut lumen and the interaction of glucose with the ileal L-cells for stimulating GLP-1 secretion. Accordingly, it appears that delaying the absorption rate of glucose to the distal ileum will modulate GLP-1 release substantially.

Therefore, a larger amount of glucose may be immediately available for absorption following rapid digestion of SUC by the brush-border sucrase in the upper small intestine, triggering a higher GIP activation. Conversely, lower degradation rate of ISO causes glucose to be slowly released, utilizing a longer portion of the small intestine for absorption (Fig. 2.5). As a result, absorption of ISO-derived glucose may be shifted to the lower intestine, thereby inducing a higher GLP-1 level. Experimental results indicate that the 1,6-glycosidic bond in ISO is cleaved not only by the brush-border isomaltase in the proximal jejunum (Dahlqvist et al. 1963; Goda and Hosoya 1983; Goda et al. 1988) but also by the brush-border glucoamylase with the highest activity in the ileum (Dahlqvist et al. 1961; Goda et al. 1988; Günther and Heymann 1998). As indicated previously, jejunal isomaltase may be saturated, hence a part of ISO may be shifted to the distal ileum for further degradation and subsequent absorption, which contributes to GLP-1 release. Enhanced GLP-1 secretion is favorable in T2DM due to significantly reduced levels but a retained insulinotropic effect, while GIP is normal or slightly elevated, but the effect is greatly reduced (Nauck et al. 1993; Vilsbøll et al. 2001).

The slower hydrolysis rate of ISO and the resulting gradual absorption of glucose in the small intestine exert influence on gastric motility and gastric emptying. With the release of GLP-1 hormone, gastric emptying is delayed (Willms et al. 1996). This regulation elicited by the small intestine is involved in the feedback control of gastric motility as an adaptation process to the intestinal digestion and absorption. When gastric emptying is slowed down, the disaccharide will move slowly to the small bowel, thereby ensuring sufficient contact with digestive enzymes and absorptive surfaces. Delayed gastric emptying may cause increased satiety, which has been observed in healthy subjects by a lower increase in plasma ghrelin levels following ISO compared with SUC consumption (van Can et al. 2009).

5.5 Plasma Glucose Turnover

5.5.1 Systemic Glucose Appearance

Total glucose entry into the systemic circulation is the sum of exogenous and endogenous glucose appearance. During the 4-h postprandial period, an average amount of 31 g of exogenous glucose originating from ISO reached the systemic circulation, with a concomitant EGP of ~21 g, yielding a total amount of ~53 g of glucose that was released into the blood circulation. Conversely, after SUC intake, a higher amount of glucose was delivered into the systemic circulation (~80 g), due to increases in both exogenous and endogenous glucose release (~41 and ~38 g, respectively). Consumption of ISO significantly attenuated postprandial oral exogenous glucose appearance, suppressed EGP, and thus improved overall postprandial glucose appearance by ~35 % compared with ingestion of rapidly absorbed SUC in T2DM subjects.

5.5.2 Oral Glucose Appearance

Subjects in the ISO/SUC-Clamp Study consumed 1 g/kg body weight of ISO or SUC. They had an average body weight of 90 kg, which means that they ingested an average amount of 90 g of ISO or SUC. Since ISO and SUC consist of 50 % glucose and 50 % fructose, glucose ingested per person is equivalent to half of the amount of oral ISO or SUC load, corresponding to an average of 45 g of glucose. Both disaccharides were enriched with ^{13}C at carbon-1 in the glucose molecules, allowing for calculation of oral glucose appearance accurately. Following ingestion of ISO or SUC containing ~45 g of glucose, an amount of ~31 or ~41 g reached the systemic circulation, respectively during the 4-h postprandial period. This corresponds to ~69 % or ~92 % of the total amount of glucose ingested by the subjects.

A variety of factors plays an important role in determining the lower magnitude of oral glucose released into the systemic circulation following the ISO load. These include rates of gastric emptying, degradation rates of disaccharides by digestive enzymes in the small intestine, absorption of glucose from the gut lumen into the portal vein, initial glucose extraction by splanchnic tissues, and finally the release of the oral absorbed glucose into the systemic circulation. Collectively, all these factors contribute substantially to the reduction of oral glucose appearance originating from ISO. Because the difference between ISO and SUC lies in the position of glycosidic bond between the two monosaccharides glucose and fructose, for which the cleavage requires different brush-border enzymes located at different sites of small intestine, it can be concluded that digestion rate of ISO represents the major determinant for the lower rates of oral glucose appearance in the systemic circulation. All other factors can be seen as accompanying factors secondary to the slower degradation rate.

5.5.3 Endogenous Glucose Production

Suppressed EGP following ISO administration can be attributed to the hormonal changes in plasma GLP-1, insulin, and glucagon concentrations. GLP-1 hormone acts to stimulate insulin secretion and inhibit glucagon release (Nauck et al. 1993; Vella et al. 2000), which explain the lower increase in plasma glucagon concentrations following ISO versus SUC consumption. GLP-1 hormone may therefore inhibit EGP indirectly through modulation of glucagon and insulin secretions. Glucagon will increase EGP preferentially via glycogenolysis (Magnusson et al. 1995), by contrast insulin exerts the opposite effect. Balance between plasma glucagon and insulin concentrations is considered a major regulator of EGP, suggesting reduced glucagon and/or increased insulin levels favor EGP suppression. Indeed, the mean molar ratio of insulin-to-glucagon concentration was increased after ISO versus SUC ingestion (18.1 ± 0.9 vs. 15.8 ± 1.0, $P < 0.001$).

Other factor, such as decreased availability of gluconeogenic precursors, may also contribute to the lower EGP following ISO as opposed to SUC consumption in T2DM patients. A reduction in plasma FFA levels was observed after administration of ISO compared with SUC in normal healthy probands and impaired glucose-tolerant subjects (van Can et al. 2009, 2012), indicating a reduced substrate supply for gluconeogenesis. Alteration in cellular metabolism may be another contributing factor to the lower endogenous glucose release. In animal studies, adaptation to dietary starch for 1 week has been shown to downregulate two key gluconeogenic enzymes (i.e., phosphoenolpyruvate carboxykinase and glucose-6-phosphatase), causing a lower glucose release in isolated hepatocytes compared to SUC-adapted diet (Bizeau et al. 2001a,b).

Numerous studies have demonstrated impaired suppression of EGP following meal ingestion or glucose administration in T2DM patients compared to healthy probands (Firth et al. 1986; Ferrannini et al. 1988; Mitrakou et al. 1990; Singhal et al. 2002; Krssak et al. 2004). Inadequate inhibition of EGP contributes to the excessive postprandial hyperglycemia in T2DM. Results of the present study indicate that postprandial EGP can be significantly reduced by intake of slowly absorbed carbohydrates.

5.5.4 Splanchnic Glucose Uptake

Newly absorbed glucose of oral ingested glucose is transported into the portal vein where a fraction is extracted by the liver and the remainder is released into the systemic circulation (Gerich 2000). Approximately 69 % or 92 % of oral glucose deriving from ingestion of ISO or SUC was recovered in the systemic circulation respectively throughout the 4-h postprandial period. Since ISO and SUC are completely digested and absorbed in the small intestine (Holub et al. 2010) and no fermentation has been reported following ingestion of the disaccharides (Kashimura et al. 1990; Tamura et al. 2004; Tonouchi et al. 2011), virtually all of the oral glucose should have been absorbed from the gut by the end of the postprandial period. Therefore, nearly 31 % or 8 % of the ingested glucose was taken up during the first pass through the splanchnic tissues following ISO or SUC consumption, respectively.

The significantly higher SGU after ISO administration could be explained by the increased insulin-to-glucagon ratio compared with SUC. Both insulin and glucagon hormones regulate the activity of hepatic glucokinase, a glucose phosphorylating enzyme responsible for controlling hepatic glucose uptake. Insulin is the primary activator, while glucagon exerts inhibitory effect (Iynedjian et al. 1995). Glucose from the portal vein enters the hepatocytes via GLUT2, subsequently is phosphorylated by glucokinase to G-6-P. Phosphorylation represents the rate-determining step for hepatic glucose uptake because GLUT2 has a lower affinity for glucose than glucokinase. Autosomal dominant defect in glucokinase activity, such as MODY, is associated with impaired hepatic glucose uptake in response to hyperglycemia and hyperinsulinemia. Evidence suggests that hepatic but not intestinal glucose uptake is impaired in T2DM, which is caused by decreased hepatic glucokinase activity (Basu et al. 2001). Thus, in the presence of increased insulin-to-glucagon concentration, hepatic glucokinase activity will be enhanced, promoting hepatic glucose uptake after ISO intake. This is supported by an animal study demonstrating decreased glucokinase activity after chronic SUC feeding for 1 week compared with starch as a complex carbohydrate (Bizeau et al. 2001b).

It is also possible that the decreased SGU observed after SUC intake may be triggered by the increased EGP. Both glycogenolysis and gluconeogenesis are involved in the process of EGP; the latter one is disproportionately high in T2DM (Magnusson et al. 1992). Failure to suppress the enhanced EGP causes continuous augmentation of intracellular glucose or G-6-P levels, which could in turn diminish SGU. Conversely, it can be expected that the decreased EGP after ISO load may promote hepatic glucose uptake. Thus, both diminished net SGU (Ludvik et al. 1997; Basu et al. 2001) and inadequate suppression of EGP as indicated above contribute to excessive and prolonged hyperglycemia in T2DM patients after glucose administration or mixed meal ingestion. Due to these impairments, a meal that attenuates postprandial hyperglycemia by inhibiting EGP and increasing SGU is preferable for T2DM subjects. Based on the results of the present study, slowly absorbed carbohydrates could help control postprandial hyperglycemia in T2DM.

5.5.5   Systemic Glucose Disappearance

During the 4-h postprandial period following ingestion of ISO or SUC containing ~45 g of glucose, an amount of ~53 or ~80 g was respectively removed from the systemic circulation. This represents a greater quantity of glucose than that ingested by the T2DM subjects. The reason for this observation is attributed to the fact that EGP was only partially suppressed following the ISO load or even increased after the SUC bolus. Of the systemic glucose disposal, ~31 or ~41 g originated from the ingested glucose and ~21 or ~38 g was derived from the EGP after the ISO or SUC load, respectively. Considering the amount of ingested glucose not reaching the systemic circulation because of the uptake by splanchnic tissues, whole-body glucose disposal will equal ~68 or ~84 g respectively following ISO or SUC ingestion (~53 or ~80 g of systemic glucose disposal plus ~15 or ~4 g of splanchnic glucose disposal). By calculating the differences indicated above, both systemic and whole-body glucose disposals were respectively ~27 and ~16 g (~35 % and ~20 %) lower after ISO versus SUC consumption.

5.6 ^{13}C-Breath Test

The results of the ^{13}C-breath test provide evidence that glucose from ISO and SUC is differently metabolized following its entry into cells. Glycolysis represents an important pathway, which degrades glucose into pyruvate in the cytosol with subsequent oxidation to CO_2 in the mitochondrial membranes and a further CO_2 formation in the citric acid cycle. As shown by the recovery of exhaled $^{13}CO_2$, one fraction of glucose from either ISO or SUC is finally oxidized to CO_2 with a lower $^{13}CO_2$ breath excretion observed after ISO intake. This can be reasonably explained by the lower whole-body and systemic glucose disposals following ISO versus SUC consumption. Because $^{13}CO_2$ abundance values did not return to baseline at the end of the 4-h postprandial period, overall breath $^{13}CO_2$ may have been underestimated. However, this underestimation can be due to incorporation of labeled ^{13}C into glycogen or gluconeogenesis. A portion of glucose may undergo non-oxidative glycolysis, leading to formation of labeled precursors (pyruvate, lactate, or alanine), which are then transformed into glucose in the liver through the Cori cycle or alanine cycle.

5.7 Insulin Sensitivity

Tissue glucose disposal can occur either by insulin-dependent or insulin-independent mechanisms. The latter component, which is determined by the mass action effect of plasma glucose concentration per se (glucose effectiveness), was assumed in the calculation of postprandial insulin action in the present study. This may possibly lead to underestimation or overestimation of S_I following administration of ISO or SUC. However, the calculated S_I value was only slightly affected by changes in glucose effectiveness. For instance, a $\pm 20\%$ deviation of the assumed value produces a maximal change of 2–5 % in the estimation of S_I. Thus, the calculation of postprandial insulin action does not largely depend on the changes in glucose effectiveness. In addition, S_I estimation by means of oral glucose minimal model provides an accurate measurement of insulin sensitivity after an oral glucose load or a meal ingestion and is as powerful as that obtained from an IVGTT (Cobelli et al. 2007).

Postprandial insulin action was significantly increased following the oral load of ISO compared with SUC, indicating that slowly absorbed carbohydrates may improve insulin resistance in T2DM subjects. With regard to the present findings on suppressed EGP and increased SGU, improvement may involve hepatic rather than peripheral insulin action after ISO administration. Systemic glucose disposal may contribute minimally to the enhanced insulin action, due to a lesser amount of oral and endogenous glucose that needed to be removed from the systemic circulation following ISO versus SUC consumption. Thus, ISO exerts a sparing effect on insulin requirement, which lowers the demand on pancreatic β-cells. Consistent with these results, a significant reduction of HOMA-IR was observed following 4 weeks consumption of ISO in hyperlipidemic subjects (Holub et al. 2010). Hence, slowly absorbed carbohydrates may ameliorate insulin sensitivity in T2DM.

Insulin action was lower following consumption of ISO+WS compared to ISO+C, suggesting that different types of ingested proteins modulate insulin action differently. Rapidly absorbed protein mixture such as whey/soy reduced insulin action to a greater extent than slowly absorbed casein. It has been reported that consumption of protein from animal sources (meat, milk/products, and cheese), but not from vegetables, is associated with a higher prevalence of diabetes (Sluijs et al. 2010; Pounis et al. 2010). Typically, animal proteins are better digested (Hoffman and Falvo 2005) and therefore may be absorbed at a faster rate than vegetable proteins. Moreover, diets rich in protein may have long-term adverse effects on glycemic control, such as impaired suppression of hepatic glucose output by insulin, promotion of insulin resistance, and increased gluconeogenesis (Linn et al. 1996, 2000). Excessive protein intake should therefore be avoided.

5.8   Glucose Infusion Rates

Following the 3-h insulin infusion prior to administration of ISO or SUC, GINF rates reached a value of $\sim$15 μmol/kg/min, indicating a considerable impairment of insulin action as opposed to the value obtained in healthy subjects under the same experimental conditions (vs. $\sim$25 μmol/kg/min) (Ang et al. 2014). As the oral glucose derived from both disaccharides was absorbed, exogenous GINF was reduced appropriately for maintaining euglycemia. Using preliminary data of a pilot study in T2DM subjects, an algorithm developed by Furler et al. (1986) has been previously applied in order to simulate the appropriate GINF reduction for maintaining euglycemia in these subjects. However, this simulation yielded unreasonable GINF rates (i.e., negative values). Therefore, plasma glucose levels increased following intake of ISO or SUC, even though the GINF rates were reduced to minimal values of $\sim$3.3 and $\sim$1.3 μmol/kg/min, respectively. These results are consistent with other studies using a combination technique of hyperinsulinemic-euglycemic clamp and oral OGTT in T2DM patients (Ludvik et al. 1997; Bajaj et al. 2003).

5.9   Plasma Amino Acid Concentrations

Plasma TAA concentrations increased higher after administration of ISO+WS compared to those following intake of ISO+C. Differences in the rise of plasma TAA levels are attributed to the faster absorption of whey compared to casein. Whey proteins are readily soluble in gastric juice, whereas casein proteins tend to clot in the stomach due to precipitation by gastric acid, thus slowing gastric emptying (Boirie et al. 1997).

Overall, postprandial individual amino acid responses were enhanced after ingestion of ISO+WS as opposed to ingestion of ISO+C. The greatest difference was seen in the aspartic acid response ($\sim$158 %), which is consistent with its content in whey/soy and casein (12.2 vs. 6.6 g/100 g). Substantial increments of plasma alanine, leucine, valine, lysine, isoleucine, and threonine were observed in the present study, confirming previous findings following ingestion of meals or drinks containing whey in healthy subjects (Hall et al. 2003; Nilsson et al. 2004; Tang et al. 2009; Veldhorst et al. 2009). The highest increase was observed in the alanine response, despite the fact that glutamic acid is the most abundant amino acid in whey/soy and casein. This could be explained by transamination of glutamic acid to alanine, which occurs during intestinal absorption, followed by the formation of further metabolic products, such as alpha-ketoglutarate, glutamine, and glutathione (Steglink et al. 1983; WHO 1988). In addition, conversion of glutamine to alanine may increase due to substantial alterations of glutamine and alanine metabolism in T2DM (Stumvoll et al. 1996).

5.10   Limitations

Several limitations of the methods used in the present ISO/SUC-Clamp Study need to be taken into consideration. Most of the T2DM patients were insulin resistant as indicated by the relatively low GINF rates prior to ISO or SUC load. Thus, following ingestion of both disaccharides, plasma glucose concentrations increased in these patients, even though the GINF was reduced to near zero. Nevertheless, application of double-tracer approach enabled accurate measurement of systemic glucose appearance and disappearance, and distinction between endogenous and exogenous glucose release into the systemic circulation.

Although the dual-isotope method pioneered by Steele et al. (1968) is reliable in a steady-state condition, calculation of glucose turnover rates are less accurate under non-steady states. Marked decreases in plasma [6,6-^{2}H$_2$]glucose TTR, particularly following SUC intake, may result in an overestimation of systemic glucose appearance. Because the Steele's 1-CM approach with a constant pV has been criticized for lacking accuracy of measuring glucose kinetics under non-steady states, a varying pV was applied yielding estimations that were closely to the results calculated with the Mari's 2-CM. The latter one is precise in measuring glucose kinetics at disequilibrium (Basu et al. 2003).

Another possible limitation is the hyperinsulinemic condition induced by insulin infusion prior to administration of ISO or SUC. Insulin infusion increased basal plasma insulin to physiological levels ($\sim$350 pmol/L), exerting action on glucose disposal and glucose production. Plasma glucose, C-peptide, and glucagon concentrations decreased respectively $\sim$30 %, $\sim$60 %, and $\sim$45 % prior to ISO or SUC load. Thus, the influence of insulin infusion on the moderate increase in plasma glucose levels and changes in the total rates of glucose kinetics following ISO intake cannot be excluded. Despite of this limitation, the present findings indicate that slowly absorbed carbohydrates are useful for controlling excessive postprandial hyperglycemia in T2DM patients. Because glucose is absorbed slowly from the gastrointestinal tract, GLP-1 secretion is triggered extensively, causing reduced glucagon release and increased insulin-to-glucagon ratio, which concomitantly shift endogenous glucose towards a lower production. Lower endogenous insulin secretion may preserve β-cell function. All of these actions contribute to improve glycemic control in individuals with T2DM.

5.11   High Protein Diets in Type 2 Diabetes Mellitus

Epidemiological studies strongly suggest that higher protein intake is associated with an increased risk for T2DM (Sluijs et al. 2010; Pounis et al. 2010; Wang et al. 2010; Tinker et al. 2011). Replacing 5 % of dietary energy derived from carbohydrate or fat with protein increased diabetes risk by $\sim$30 % in the EPIC-NL cohort during a 10-year follow-up of 38,094 participants (Sluijs et al. 2010).

In another study, a 20 % higher protein consumption was associated with an 82 % higher risk of diabetes in a total of 74,155 postmenopausal women (Tinker et al. 2011). In the Insulin Resistance Atherosclerosis Study, dietary pattern was found to be strongly associated with the risk of developing T2DM in 880 middle-aged participants over a 5-year follow-up, independent of age, sex, race/ethnicity/clinic, family history of diabetes, basal glucose tolerance status, energy expenditure, smoking, and energy intake. Dietary pattern characterized by higher intakes of various proteins, such as red meat, low-fiber bread and cereal, dried beans, eggs, cheese, and cottage cheese, increased the odds of developing diabetes by more than 4-fold (Liese et al. 2009). Thus, these studies emphasize the importance of monitoring and avoiding excessive protein intake in the prevention, reduction, and management of T2DM.

Dietary protein intake in diabetic patients usually ranges between 15–20 % of total daily energy in most industrialized countries, which corresponds to 1.3–2.0 g/kg body weight and represents an intake that exceeds the recommended daily amount of 0.8 g/kg body weight. Currently, there are no guidelines addressing protein quality for the management of diabetes (Mann et al. 2004; Evert et al. 2014). Present findings suggest that ingestion of fast absorbed protein mixtures may not be recommended for glycemic control of T2DM patients. Long-term effects of the consumption of proteins with different absorption rates remain to be investigated.

5.12   Benefits of Low Glycemic Index Diet

Slowly absorbed ISO or addition of ISO to both proteins increased plasma glucose and insulin levels moderately. Incremental peak glucose and insulin levels were higher in previous studies with diabetic patients (Gannon et al. 1988, 1992; van Loon et al. 2003; Manders et al. 2005, 2006) but were similar after a lunch meal containing mashed potatoes and whey (Frid et al. 2005) compared to the present study (~5–10 vs. ~3.5 mmol/L and ~500–1700 vs. ~120–380 pmol/L, respectively). Since the amounts of ingested carbohydrate (~45–60 g) and protein (~25–30 g) were comparable (Gannon et al. 1988, 1992; Manders et al. 2006; Frid et al. 2005), part of the variation in publications may be explained by the different carbohydrate sources (glucose, maltodextrin, or mashed potatoes). Altogether, modulating the type of carbohydrates in the diabetic diet is effective in controlling postprandial hyperglycemia. Results of meta-analyses support the use of low GI foods in the metabolic control of diabetic patients (Brand-Miller et al. 2003; Thomas and Elliott 2010).

Daily regular consumption of 50 g ISO over 12 weeks did not affect glycemic control assessed as HbA_{1c} but significantly lowered plasma triglyceride levels in free-living T2DM patients in addition to their standard antidiabetic treatment as opposed to SUC consumption (Brunner et al. 2012). As indicated by the authors of the study, ISO and SUC constituted only ~10 % of the total calorie intake, representing an overall reduction of GI by ~6 units, which was obviously minor to evoke distinct effects on metabolic control within a mixed diet. Additionally, results of the combined uptake of carbohydrates with proteins indicate that dietary proteins modify insulin and glucose responses, and may therefore affect glycemic control as well.

Long-term studies have shown some beneficial effects of a low GI diet compared with a high GI diet in subjects with IGT and T2DM. These include lower glucose and insulin responses, improved lipid profiles and capacity for fibrinolysis, as well as increased glucose disposal in T2DM patients consuming similar energy intake and macronutrient composition (Järvi et al. 1999; Rizkalla et al. 2004). In obese subjects with IGT, regular low GI diet has been reported to increase disposition index, which is known as an index for β-cell function (Wolever and Mehling 2002), and ameliorate pancreatic β-cell and intestinal K-cell functions (Solomon et al. 2010).

5.13   Conclusions and Outlook

The results of this work provide novel and comprehensive details on the effects of ISO consumption on postprandial glucose homeostasis in T2DM. ISO is completely but absorbed significantly slower into the blood circulation as opposed to SUC. This is associated with a greater release of GLP-1 hormone, which inhibits glucagon secretion and shifts toward increased insulin-to-glucagon ratio. Consequently, ingestion of ISO attenuates oral and endogenous glucose release into the systemic circulation, enhances first-pass glucose extraction by splanchnic tissues, and thus decreases overall postprandial glucose flux compared with ingestion of SUC in T2DM subjects. Insulin action is enhanced concomitant to inhibition of EGP and elevation of first-pass SGU. Because of a lower amount of glucose entering the systemic circulation, a lesser quantity of glucose needs to be disposed from the circulation following ISO versus SUC intake.

Based on the results above, it can be concluded that three important factors affect the magnitude of postprandial glucose excursions: oral glucose appearance, endogenous glucose release, and initial SGU. The extent of these factors can be substantially improved by consumption of slowly as opposed to rapidly absorbed carbohydrates in T2DM.

Ingestion of ISO combined either with whey/soy or casein elevates postprandial insulin levels, but reduces insulin action, and therefore does not improve plasma glucose concentrations compared to ingestion of ISO only in T2DM subjects. Whey/soy proteins increase plasma insulin concentrations, which are associated with enhanced plasma amino acid levels, but concomitantly decrease insulin action compared to casein protein. Thus, different types of proteins modulate insulin response and insulin action differently. A fast-absorbing protein reduces insulin action to a greater extent than a slow-absorbing protein, indicating the importance of monitoring excessive protein intake, as well as the types of proteins, for glycemic control of T2DM patients. However, in recent years, low-carbohydrate high-protein diet approaches have gained wide popularity for weight loss in T2DM (Larsen et al. 2011).

Taken together, these results suggest that modulating the types of carbohydrates is effective for controlling postprandial hyperglycemia in T2DM. Therefore, slowly absorbed carbohydrates should be included in the dietary management of T2DM individuals. To broaden understanding of the mechanisms of action of slowly absorbed carbohydrates on postprandial glucose homeostasis, future research should focus on characterizing the route of postprandial glucose disposal including glycolysis, gluconeogenesis, and glycogen formation in T2DM because knowledge in this area is scarce. It is of interest to explore if other slowly absorbed carbohydrates also exert similar effects on postprandial glucose homeostasis. Future research using other carbohydrate sources will be encouraged. Finally, long-term studies are important and necessary to confirm those findings.

Chapter 6
Summary

T2DM is characterized by chronic fasting and postprandial hyperglycemia, which result from impaired insulin secretion and diminished insulin action. Epidemiological evidence suggests that controlling postprandial plasma glucose excursions is important in order to reduce the development of long-term cardiovascular complications in T2DM individuals. A low GI diet has been reported to improve glycemic control effectively. The GI measures the extent to which carbohydrates affect blood glucose. Carbohydrates with a low GI are usually slowly absorbed, producing delayed gradual increases in blood glucose and insulin levels. For example, ISO, an isomer of SUC, is digested slower than other sugars such as SUC or maltose. Consumption of ISO attenuates postprandial glucose and insulin levels compared with SUC in healthy, impaired glucose-tolerant, and T2DM subjects. The mechanisms by which ISO limits postprandial hyperglycemia have not been clarified. Besides, there is evidence to suggest that a combined uptake of carbohydrates with proteins triggers additional insulin secretion. However, results have been inconclusive in regard to the effects of insulin stimulation on glucose homeostasis in T2DM subjects. It remains unclear whether a combined load of protein and slowly digested ISO could improve postprandial glucose response in these subjects.

The objectives of this work were therefore to characterize the mechanisms and effects of ISO administration on postprandial glucose metabolism compared with SUC in T2DM subjects and to investigate the effects of co-ingestion of ISO and proteins on glucose and insulin responses in these subjects.

Two short-term studies with randomized, double-blinded, and crossover design were applied. In the ISO/SUC-Clamp Study, eleven T2DM subjects underwent initially a 3-h euglycemic-hyperinsulinemic (0.8 mU/kg/min) clamp that was subsequently combined with 1 g/kg body weight of an oral ^{13}C-enriched ISO or SUC load. Hormonal responses and glucose kinetics were analyzed during a 4-h postprandial period. In the ISO-Protein Study, thirty T2DM subjects consumed 50 g of ISO combined either with a mixture of 21 g whey/soy or with 21 g casein on

© The Author 2016

M. Ang, *Metabolic Response of Slowly Absorbed Carbohydrates in Type 2 Diabetes Mellitus*, SpringerBriefs in Systems Biology, DOI 10.1007/978-3-319-27898-8_6

separate days. In another experiment, the subjects consumed only 50 g ISO. Plasma glucose and insulin responses were measured over a 4-h postprandial period.

The results of the ISO/SUC-Clamp Study showed that absorption of ISO was prolonged by $\sim$50 min compared with SUC ($P = 0.005$). Plasma responses of insulin, C-peptide, glucagon, and GIP were $\sim$50–80 % lower ($P < 0.01$); by contrast, GLP-1 was $\sim$170 % higher following administration of ISO ($P < 0.001$), which resulted in an increase in insulin-to-glucagon ratio compared with SUC ($P < 0.001$). Cumulative amount of systemic glucose appearance was $\sim$35 % lower after ISO versus SUC consumption ($P = 0.003$), due to reduction of oral and endogenous glucose release ($P < 0.001$) and a higher first-pass SGU ($P = 0.003$). Insulin action was enhanced after intake of ISO compared with SUC ($P = 0.011$). In the ISO-Protein Study, no significant differences in plasma glucose responses were observed after ingestion of ISO alone, ISO+WS or ISO+C ($P = 0.239$). Compared to ingestion of ISO alone, insulin response was $\sim$190–270 % higher ($P < 0.001$), whereas insulin action was lower after ingestion of ISO+WS or ISO+C ($P < 0.01$).

In summary, ingestion of slowly absorbed ISO significantly attenuated postprandial oral and endogenous glucose release, increased SGU, and thus reduced overall postprandial glucose flux compared with ingestion of rapidly absorbed SUC in T2DM subjects. The enhanced GLP-1 secretion observed after ISO ingestion caused inhibition of glucagon release, resulting in enhancement of insulin-to-glucagon ratio, which in turn increased insulin action concomitant to suppression of endogenous glucose release and elevation of SGU. Ingestion of ISO+WS or ISO+C elevated postprandial insulin levels, but reduced insulin action, and therefore did not improve plasma glucose concentrations compared to ingestion of ISO only in T2DM subjects. Together, these results suggest that modulating the types of carbohydrates is effective for controlling postprandial hyperglycemia in T2DM. Therefore, slowly absorbed carbohydrates should be included in the dietary management of individuals with T2DM.

Appendix A
Mathematical Analysis

This appendix describes the mathematical derivation of insulin sensitivity index S_I. The derivation is done based on the general formulation of oral glucose minimal model as introduced by Caumo et al. (2000) and Dalla Man et al. (2002). According to Caumo et al. and Dalla Man et al., the glucose changes over time $\dot{G}(t)$ and the changes of insulin action over time $\dot{X}(t)$ are defined as follows

$$\begin{cases} \dot{G}(t) = -[p_1 + X(t)] \cdot G(t) + p_1 \cdot G_b + \frac{R_a O(t)}{V}; & G(0) = G_b \\ \dot{X}(t) = -p_2 \cdot X(t) + p_3 \cdot [I(t) - I_b]; & X(0) = 0 \end{cases} \tag{A.1}$$

Both $\dot{G}(t)$ and $\dot{X}(t)$ represent the OGTT model as a dynamical system. The model can be solved mathematically with the help of Ordinary Differential Equation solver (ODE-solver) that is provided by the Maple software. The solution for $X(t)$ is described as follows

$$X(t) = C \cdot e^{-p_2 t} - \frac{p_3 \cdot [I_b - I(t)]}{p_2} \tag{A.2}$$

The C value can be calculated by applying the initial values of $X(t)$ and $I(t)$, which are $X(0) = 0$ and $I(0) = I_b$. Applying those values to Eq. A.2 yields a C value of 0. Therefore, the equation can be re-written as follows

$$X(t) = -\frac{p_3 \cdot [I_b - I(t)]}{p_2} \tag{A.3}$$

According to Bergman et al. (1979) and Dalla Man et al. (2002), S_I is calculated as the ratio of p_3 to p_2 and is written as

$$S_I = \frac{p_3}{p_2} \tag{A.4}$$

© The Author 2016

M. Ang, *Metabolic Response of Slowly Absorbed Carbohydrates in Type 2 Diabetes Mellitus*, SpringerBriefs in Systems Biology, DOI 10.1007/978-3-319-27898-8

The ratio p_3/p_2 in Eq. A.3 can be substituted by S_I according to Eq. A.4. Thus, $X(t)$ can be calculated as follows

$$X(t) = S_I(t) \cdot [I(t) - I_b] \tag{A.5}$$

Applying $X(t)$ from the Eq. A.5 to $\dot{G}(t)$ of the Eq. A.1 yields

$$\dot{G}(t) = -[p_1 + S_I(t) \cdot [I(t) - I_b]] \cdot G(t) + p_1 \cdot G_b + \frac{R_aO(t)}{V} \tag{A.6}$$

Eq. A.6 is then derived mathematically step-by-step as follows

$$\dot{G}(t) = -[p_1 + S_I(t) \cdot [I(t) - I_b]] \cdot G(t) + p_1 \cdot G_b + \frac{R_aO(t)}{V}$$

$$[p_1 + S_I(t) \cdot [I(t) - I_b]] \cdot G(t) = -\dot{G}(t) + p_1 \cdot G_b + \frac{R_aO(t)}{V}$$

$$p_1 + S_I(t) \cdot [I(t) - I_b] = \frac{\frac{R_aO(t)}{V} + p_1 \cdot G_b - \dot{G}(t)}{G(t)}$$

$$S_I(t) \cdot [I(t) - I_b] = \frac{\frac{R_aO(t)}{V} + p_1 \cdot G_b - \dot{G}(t)}{G(t)} - p_1$$

Finally, S_I is described mathematically as functions of $I(t)$, $G(t)$, $\dot{G}(t)$, and $R_aO(t)$. S_I is then defined as

$$S_I(t) = \frac{1}{I(t) - I_b} \left[\frac{\frac{R_aO(t)}{V} + p_1 \cdot G_b - \dot{G}(t)}{G(t)} - p_1 \right] \tag{A.7}$$

References

ADA (2010) Diagnosis and classification of diabetes mellitus. Diabetes Care 33(Suppl 1):S62–S69. doi:10.2337/dc10-S062

Ang M, Linn T (2014) Comparison of the effects of slowly and rapidly absorbed carbohydrates on postprandial glucose metabolism in type 2 diabetes mellitus patients: a randomized trial. Am J Clin Nutr 100(4):1059–1068. doi:10.3945/ajcn.113.076638

Ang M, Müller AS, Wagenlehner F, Pilatz A, Linn T (2012) Combining protein and carbohydrate increases postprandial insulin levels but does not improve glucose response in patients with type 2 diabetes. Metabolism 61(12):1696–1702. doi:10.1016/j.metabol.2012.05.008

Ang M, Meyer C, Brendel MD, Bretzel RG, Linn T (2014) Magnitude and mechanisms of glucose counterregulation following islet transplantation in patients with type 1 diabetes suffering from severe hypoglycaemic episodes. Diabetologia 57(3):623–632. doi:10.1007/s00125-013-3120-9

Aronoff SL, Berkowitz K, Shreiner B, Want L (2004) Glucose metabolism and regulation: beyond insulin and glucagon. Diabetes Spectr 17(3):183–190. doi:10.2337/diaspect.17.3.183

Ascaso JF, Pardo S, Real JT, Lorente RI, Priego A, Carmena R (2003) Diagnosing insulin resistance by simple quantitative methods in subjects with normal glucose metabolism. Diabetes Care 26(12):3320–3325

Atkinson FS, Foster-Powell K, Brand-Miller JC (2008) International tables of glycemic index and glycemic load values: 2008. Diabetes Care 31(12):2281–2283. doi:10.2337/dc08-1239

Atkinson MA, Eisenbarth GS, Michels AW (2014) Type 1 diabetes. Lancet 383(9911):69–82. doi:10.1016/S0140-6736(13)60591-7

Bajaj M, Suraamornkul S, Pratipanawatr T, Hardies LJ, Pratipanawatr W, Glass L, Cersosimo E, Miyazaki Y, DeFronzo RA (2003) Pioglitazone reduces hepatic fat content and augments splanchnic glucose uptake in patients with type 2 diabetes. Diabetes 52(6):1364–1370

Basu A, Basu R, Shah P, Vella A, Johnson CM, Jensen M, Nair KS, Schwenk WF, Rizza RA (2001) Type 2 diabetes impairs splanchnic uptake of glucose but does not alter intestinal glucose absorption during enteral glucose feeding: additional evidence for a defect in hepatic glucokinase activity. Diabetes 50(6):1351–1362

Basu R, Di Camillo B, Toffolo G, Basu A, Shah P, Vella A, Rizza R, Cobelli C (2003) Use of a novel triple-tracer approach to assess postprandial glucose metabolism. Am J Physiol Endocrinol Metab 284(1):E55–E69. doi:10.1152/ajpendo.00190.2001

Bayer HealthCare (2003) Insulin (IRI), Advia Centaur Testanleitung

Bayer HealthCare (2006) Glukose-Hexokinase-II (GLUH), Advia Chemistry Systems

Bergman RN, Ider YZ, Bowden CR, Cobelli C (1979) Quantitative estimation of insulin sensitivity. Am J Physiol 236(6):E667–E677

© The Author 2016

M. Ang, *Metabolic Response of Slowly Absorbed Carbohydrates in Type 2 Diabetes Mellitus*, SpringerBriefs in Systems Biology, DOI 10.1007/978-3-319-27898-8

Bergman RN, Finegood DT, Ader M (1985) Assessment of insulin sensitivity in vivo. Endocr Rev 6(1):45–86

Best JD, Taborsky GJ, Halter JB, Porte D (1981) Glucose disposal is not proportional to plasma glucose level in man. Diabetes 30(10):847–850

Best JD, Kahn SE, Ader M, Watanabe RM, Ni TC, Bergman RN (1996) Role of glucose effectiveness in the determination of glucose tolerance. Diabetes Care 19(9):1018–1030

Bizeau ME, Thresher JS, Pagliassotti MJ (2001a) A high-sucrose diet increases gluconeogenic capacity in isolated periportal and perivenous rat hepatocytes. Am J Physiol Endocrinol Metab 280(5):E695–E702

Bizeau ME, Thresher JS, Pagliassotti MJ (2001b) Sucrose diets increase glucose-6-phosphatase and glucose release and decrease glucokinase in hepatocytes. J Appl Physiol 91(5):2041–2046

Boirie Y, Dangin M, Gachon P, Vasson MP, Maubois JL, Beaufrere B (1997) Slow and fast dietary proteins differently modulate postprandial protein accretion. Proc Natl Acad Sci U S A 94(26):14930–14935

Bracken RM, Page R, Gray B, Kilduff LP, West DJ, Stephens JW, Bain SC (2012) Isomaltulose improves glycemia and maintains run performance in type 1 diabetes. Med Sci Sports Exerc 44(5):800–808. doi:10.1249/MSS.0b013e31823f6557

Bradley DC, Steil GM, Bergman RN (1993) Quantitation of measurement error with Optimal Segments: basis for adaptive time course smoothing. Am J Physiol 264(6 Pt 1):E902–E911

Brand WA (2004) Mass spectrometer hardware for analyzing stable isotope ratios. In: de Groot PA (ed) Handbook of stable isotope analytical techniques, vol 1, chap 38. Elsevier B.V., Amsterdam

Brand-Miller J, Hayne S, Petocz P, Colagiuri S (2003) Low-glycemic index diets in the management of diabetes: a meta-analysis of randomized controlled trials. Diabetes Care 26(8):2261–2267

Brunner S, Holub I, Theis S, Gostner A, Melcher R, Wolf P, Amann-Gassner U, Scheppach W, Hauner H (2012) Metabolic effects of replacing sucrose by isomaltulose in subjects with type 2 diabetes: a randomized double-blind trial. Diabetes Care 35(6):1249–1251. doi:10.2337/dc11-1485

Calbet JA, MacLean DA (2002) Plasma glucagon and insulin responses depend on the rate of appearance of amino acids after ingestion of different protein solutions in humans. J Nutr 132(8):2174–2182

Campbell PJ, Mandarino LJ, Gerich JE (1988) Quantification of the relative impairment in actions of insulin on hepatic glucose production and peripheral glucose uptake in non-insulin-dependent diabetes mellitus. Metabolism 37(1):15–21

Caumo A, Bergman RN, Cobelli C (2000) Insulin sensitivity from meal tolerance tests in normal subjects: a minimal model index. J Clin Endocrinol Metab 85(11):4396–4402. doi:10.1210/jcem.85.11.6982

Chaikomin R, Wu KL, Doran S, Jones KL, Smout AJPM, Renooij W, Holloway RH, Meyer JH, Horowitz M, Rayner CK (2007) Concurrent duodenal manometric and impedance recording to evaluate the effects of hyoscine on motility and flow events, glucose absorption, and incretin release. Am J Physiol Gastrointest Liver Physiol 292(4):G1099–G1104. doi:10.1152/ajpgi.00519.2006

Cobelli C, Mari A, Ferrannini E (1987) Non-steady state: error analysis of Steele's model and developments for glucose kinetics. Am J Physiol 252(5 Pt 1):E679–E689

Cobelli C, Toffolo G, Foster DM (1992) Tracer-to-tracee ratio for analysis of stable isotope tracer data: link with radioactive kinetic formalism. Am J Physiol 262(6 Pt 1):E968–E975

Cobelli C, Toffolo GM, Dalla Man C, Campioni M, Denti P, Caumo A, Butler P, Rizza R (2007) Assessment of beta-cell function in humans, simultaneously with insulin sensitivity and hepatic extraction, from intravenous and oral glucose tests. Am J Physiol Endocrinol Metab 293(1):E1–E15

Coggan AR (1999) Use of stable isotopes to study carbohydrate and fat metabolism at the whole-body level. Proc Nutr Soc 58(4):953–961

Consoli A (1992) Role of liver in pathophysiology of NIDDM. Diabetes Care 15(3):430–441

Consoli A, Nurjahan N, Gerich JE, Mandarino LJ (1992) Skeletal muscle is a major site of lactate uptake and release during hyperinsulinemia. Metabolism 41(2):176–179

Cryer PE (2008) The barrier of hypoglycemia in diabetes. Diabetes 57(12):3169–3176

Dahlqvist A, Raunio R, Sjöberg B, Dam H, Toft J (1961) Hydrolysis of palatinose (isomaltulose) by pig intestinal glycosidases. Acta Chem Scand 15:808–816. doi:10.3891/acta.chem.scand. 15-0808

Dahlqvist A, Auricchio S, Semenza G, Prader A (1963) Human intestinal disaccharidases and hereditary disaccharide intolerance. The hydrolysis of sucrose, isomaltose, palatinose (isomaltulose), and a 1,6-alpha-oligosaccharide (isomalto-oligosaccharide) preparation. J Clin Invest 42:556–562. doi:10.1172/JCI104744

Dako (2009) C-peptide ELISA kit, Code K6220

Dalla Man C, Caumo A, Cobelli C (2002) The oral glucose minimal model: estimation of insulin sensitivity from a meal test. IEEE Trans Biomed Eng 49(5):419–429. doi:10.1109/10.995680

Dalla Man C, Caumo A, Basu R, Rizza R, Toffolo G, Cobelli C (2004) Minimal model estimation of glucose absorption and insulin sensitivity from oral test: validation with a tracer method. Am J Physiol Endocrinol Metab 287(4):E637–E643. doi:10.1152/ajpendo.00319.2003

Dalla Man C, Yarasheski KE, Caumo A, Robertson H, Toffolo G, Polonsky KS, Cobelli C (2005) Insulin sensitivity by oral glucose minimal models: validation against clamp. Am J Physiol Endocrinol Metab 289(6):E954–E959. doi:10.1152/ajpendo.00076.2005

Danaei G, Finucane MM, Lu Y, Singh GM, Cowan MJ, Paciorek CJ, Lin JK, Farzadfar F, Khang YH, Stevens GA, Rao M, Ali MK, Riley LM, Robinson CA, Ezzati M (2011) National, regional, and global trends in fasting plasma glucose and diabetes prevalence since 1980: systematic analysis of health examination surveys and epidemiological studies with 370 country-years and 2.7 million participants. Lancet 378(9785):31–40. doi:10.1016/ S0140-6736(11)60679-X

Debodo RC, Steele R, Altszuler N, Dunn A, Bishop JS (1963) On the hormonal regulation of carbohydrate metabolism; studies with C14 glucose. Recent Prog Horm Res 19:445–488

DECODE Study Group (2001) Glucose tolerance and cardiovascular mortality: comparison of fasting and 2-hour diagnostic criteria. Arch Intern Med 161(3):397–405

DECODE Study Group (2003) Is the current definition for diabetes relevant to mortality risk from all causes and cardiovascular and noncardiovascular diseases? Diabetes Care 26(3):688–696

DeFronzo RA (2004) Pathogenesis of type 2 diabetes mellitus. Med Clin North Am 88(4):787–835. doi:10.1016/j.mcna.2004.04.013

DeFronzo RA, Tobin JD, Andres R (1979) Glucose clamp technique: a method for quantifying insulin secretion and resistance. Am J Physiol Endocrinol Metab 237(3):E214–E223

Del Prato S (2002) In search of normoglycaemia in diabetes: controlling postprandial glucose. Int J Obes Relat Metab Disord 26(Suppl 3):S9–S17. doi:10.1038/sj.ijo.0802172

Eelderink C, Schepers M, Preston T, Vonk RJ, Oudhuis L, Priebe MG (2012) Slowly and rapidly digestible starchy foods can elicit a similar glycemic response because of differential tissue glucose uptake in healthy men. Am J Clin Nutr 96(5):1017–1024. doi:10.3945/ajcn.112.041947

Ehrlein HJ, Schemann M (2005) Gastrointestinal motility. http://humanbiology.wzw.tum.de/ fileadmin/Bilder/tutorials/tutorial.pdf. Last visited 01 June 2014

EU Commision (2005) Commission Decision of 25th July 2005 authorising the placing on the market of isomaltulose as a novel food or novel food ingredient under Regulation (EC) no 258/97 of the European Parliament and of the Council. Official Journal of the European Union L199/90–L199/91

Evert AB, Boucher JL, Cypress M, Dunbar SA, Franz MJ, Mayer-Davis EJ, Neumiller JJ, Nwankwo R, Verdi CL, Urbanski P, Yancy WS (2014) Nutrition therapy recommendations for the management of adults with diabetes. Diabetes Care 37(Suppl 1):S120–S143. doi: 10.2337/dc14-S120

FDA (2006) U.S. GRAS Notification no. 0184 on isomaltulose.

Fekkes D, van Dalen A, Edelman M, Voskuilen A (1995) Validation of the determination of amino acids in plasma by high-performance liquid chromatography using automated pre-column derivatization with o-phthaldialdehyde. J Chromatogr B Biomed Appl 669(2):177–186

Ferrannini E, Simonson DC, Katz LD, Reichard G, Bevilacqua S, Barrett EJ, Olsson M, DeFronzo RA (1988) The disposal of an oral glucose load in patients with non-insulin-dependent diabetes. Metabolism 37(1):79–85

Finegood DT, Bergman RN (1983) Optimal segments: a method for smoothing tracer data to calculate metabolic fluxes. Am J Physiol 244(5):E472–E479

Finegood DT, Pacini G, Bergman RN (1984) The insulin sensitivity index. Correlation in dogs between values determined from the intravenous glucose tolerance test and the euglycemic glucose clamp. Diabetes 33(4):362–368

Finegood DT, Bergman RN, Vranic M (1987) Estimation of endogenous glucose production during hyperinsulinemic-euglycemic glucose clamps. Comparison of unlabeled and labeled exogenous glucose infusates. Diabetes 36(8):914–924

Firth RG, Bell PM, Marsh HM, Hansen I, Rizza RA (1986) Postprandial hyperglycemia in patients with noninsulin-dependent diabetes mellitus. Role of hepatic and extrahepatic tissues. J Clin Invest 77(5):1525–1532. doi:10.1172/JCI112467

Frid AH, Nilsson M, Holst JJ, Bjorck IM (2005) Effect of whey on blood glucose and insulin responses to composite breakfast and lunch meals in type 2 diabetic subjects. Am J Clin Nutr 82(1):69–75

Furler SM, Zelenka GS, Kraegen EW (1986) Development and testing of a simple algorithm for a glucose clamp. Med Biol Eng Comput 24(4):365–370

Gannon MC, Nuttall FQ, Neil BJ, Westphal SA (1988) The insulin and glucose responses to meals of glucose plus various proteins in type II diabetic subjects. Metabolism 37(11):1081–1088

Gannon MC, Nuttall FQ, Lane JT, Burmeister LA (1992) Metabolic response to cottage cheese or egg white protein, with or without glucose, in type II diabetic subjects. Metabolism 41(10):1137–1145

Genuth S, Alberti KGMM, Bennett P, Buse J, Defronzo R, Kahn R, Kitzmiller J, Knowler WC, Lebovitz H, Lernmark A, Nathan D, Palmer J, Rizza R, Saudek C, Shaw J, Steffes M, Stern M, Tuomilehto J, Zimmet P (2003) Follow-up report on the diagnosis of diabetes mellitus. Diabetes Care 26(11):3160–3167

Gerich JE (1993) Control of glycaemia. Baillieres Clin Endocrinol Metab 7(3):551–586

Gerich JE (2000) Physiology of glucose homeostasis. Diabetes Obes Metab 2(6):345–350

Goda T, Hosoya N (1983) Hydrolysis of palatinose by rat intestinal sucrase-isomaltase complex. Nihon Eiyo Shokuryo Gakkai Shi 36:169–173

Goda T, Takase S, Hosoya N (1988) Hydrolysis of alpha-D-glucopyranosyl-1,6-sorbitol and alpha-D-glucopyranosyl-1,6-mannitol by rat intestinal disaccharidases. J Nutr Sci Vitaminol 34(1):131–140

Goodman BE (2010) Insights into digestion and absorption of major nutrients in humans. Adv Physiol Educ 34(2):44–53. doi:10.1152/advan.00094.2009

Gregersen Sr, Jeppesen PB, Holst JJ, Hermansen K (2004) Antihyperglycemic effects of stevioside in type 2 diabetic subjects. Metabolism 53(1):73–76

Günther S, Heymann H (1998) Di- and oligosaccharide substrate specificities and subsite binding energies of pig intestinal glucoamylase-maltase. Arch Biochem Biophys 354(1):111–116. doi:10.1006/abbi.1998.0684

Gutniak M, Orskov C, Holst JJ, Ahrén B, Efendic S (1992) Antidiabetogenic effect of glucagon-like peptide-1 (7-36)amide in normal subjects and patients with diabetes mellitus. N Engl J Med 326(20):1316–1322. doi:10.1056/NEJM199205143262003

Häberer D, Thibault L, Langhans W, Geary N (2009) Beneficial effects on glucose metabolism of chronic feeding of isomaltulose versus sucrose in rats. Ann Nutr Metab 54(1):75–82. doi:10.1159/000207358

Hageman R, Severijnen C, van de Heijning BJ, Bouritius H, van Wijk N, van Laere K, van der Beek EM (2008) A specific blend of intact protein rich in aspartate has strong postprandial glucose attenuating properties in rats. J Nutr 138(9):1634–1640

Hall EJ, Batt RM (1996) Urinary excretion by dogs of intravenously administered simple sugars. Res Vet Sci 60(3):280–282

Hall WL, Millward DJ, Long SJ, Morgan LM (2003) Casein and whey exert different effects on plasma amino acid profiles, gastrointestinal hormone secretion and appetite. Br J Nutr 89(2):239–248

Herrmann C, Göke R, Richter G, Fehmann HC, Arnold R, Göke B (1995) Glucagon-like peptide-1 and glucose-dependent insulin-releasing polypeptide plasma levels in response to nutrients. Digestion 56(2):117–126

Hira T, Muramatsu M, Okuno M, Hara H (2011) GLP-1 secretion in response to oral and luminal palatinose (isomaltulose) in rats. J Nutr Sci Vitaminol 57(1):30–35

Hoffman JR, Falvo MJ (2005) Protein – Which is best ? Nutrition 3(2004):118–130

Holst JJ, Gromada J (2004) Role of incretin hormones in the regulation of insulin secretion in diabetic and nondiabetic humans. Am J Physiol Endocrinol Metab 287(2):E199–E206. doi: 10.1152/ajpendo.00545.2003

Holub I, Gostner A, Theis S, Nosek L, Kudlich T, Melcher R, Scheppach W (2010) Novel findings on the metabolic effects of the low glycaemic carbohydrate isomaltulose (Palatinose). Br J Nutr 103(12):1730–1737. doi:10.1017/S0007114509993874

Horowitz M, Edelbroek MA, Wishart JM, Straathof JW (1993) Relationship between oral glucose tolerance and gastric emptying in normal healthy subjects. Diabetologia 36(9):857–862

Hosker JP, Matthews DR, Rudenski AS, Burnett MA, Darling P, Bown EG, Turner RC (1985) Continuous infusion of glucose with model assessment: measurement of insulin resistance and beta-cell function in man. Diabetologia 28(7):401–411

IDF (2014) Guideline for management of postmeal glucose in diabetes. Diabetes Res Clin Pract 103(2):256–268. doi:10.1016/j.diabres.2012.08.002

Insel PA, Liljenquist JE, Tobin JD, Sherwin RS, Watkins P, Andres R, Berman M (1975) Insulin control of glucose metabolism in man: a new kinetic analysis. J Clin Invest 55(5):1057–1066. doi:10.1172/JCI108006

International Expert Committee (2009) International Expert Committee report on the role of the A1C assay in the diagnosis of diabetes. Diabetes Care 32(7):1327–1334. doi:10.2337/dc09-9033

Iynedjian PB, Marie S, Gjinovci A, Genin B, Deng SP, Buhler L, Morel P, Mentha G (1995) Glucokinase and cytosolic phosphoenolpyruvate carboxykinase (GTP) in the human liver. Regulation of gene expression in cultured hepatocytes. J Clin Invest 95(5):1966–1973. doi: 10.1172/JCI117880

Järvi AE, Karlström BE, Granfeldt YE, Björck IE, Asp NG, Vessby BO (1999) Improved glycemic control and lipid profile and normalized fibrinolytic activity on a low-glycemic index diet in type 2 diabetic patients. Diabetes Care 22(1):10–18

Jenkins DJ, Wolever TM, Taylor RH, Barker H, Fielden H, Baldwin JM, Bowling AC, Newman HC, Jenkins AL, Goff DV (1981) Glycemic index of foods: a physiological basis for carbohydrate exchange. Am J Clin Nutr 34(3):362–366

Jenkins DJ, Wolever TM, Jenkins AL, Taylor RH (1987) Dietary fibre, carbohydrate metabolism and diabetes. Mol Aspects Med 9(1):97–112

Jenkins DJA, Kendall CWC, Augustin LSA, Franceschi S, Hamidi M, Marchie A, Jenkins AL, Axelsen M (2002) Glycemic index: overview of implications in health and disease. Am J Clin Nutr 76(1):266S–273S

Kashimura J, Nakajima Y, Benno Y, Mitsuoka T (1990) Comparison of fecal microflora among subjects given palatinose and its condensates. Nihon Eiyo Shokuryo Gakkai Shi 43:175–180

Katz J, Dunn A, Chenoweth M, Golden S (1974) Determination of synthesis, recycling and body mass of glucose in rats and rabbits in vivo 3H- and 14C-labelled glucose. Biochem J 142(1):171–183

Kawai K, Okuda Y, Yamashita K (1985) Changes in blood glucose and insulin after an oral palatinose administration in normal subjects. Endocrinol Jpn 32(6):933–936

Kawai K, Okuda Y, Chiba Y, Yamashita K (1986) Palatinose as a potential parenteral nutrient: its metabolic effects and fate after oral and intravenous administration to dogs. J Nutr Sci Vitaminol 32(3):297–306

Kawai K, Yoshikawa H, Murayama Y, Okuda Y, Yamashita K (1989) Usefulness of palatinose as a caloric sweetener for diabetic patients. Horm Metab Res 21(6):338–340. doi:10.1055/s-2007-1009230

Kim W, Egan JM (2008) The role of incretins in glucose homeostasis and diabetes treatment. Pharmacol Rev 60(4):470–512. doi:10.1124/pr.108.000604

Krebs M, Krssak M, Bernroider E, Anderwald C, Brehm A, Meyerspeer M, Nowotny P, Roth E, Waldhausl W, Roden M (2002) Mechanism of amino acid-induced skeletal muscle insulin resistance in humans. Diabetes 51(3):599–605

Krebs M, Brehm A, Krssak M, Anderwald C, Bernroider E, Nowotny P, Roth E, Chandramouli V, Landau BR, Waldhausl W, Roden M (2003) Direct and indirect effects of amino acids on hepatic glucose metabolism in humans. Diabetologia 46(7):917–925

Krssak M, Brehm A, Bernroider E, Anderwald C, Nowotny P, Dalla Man C, Cobelli C, Cline GW, Shulman GI, Waldhäusl W, Roden M (2004) Alterations in postprandial hepatic glycogen metabolism in type 2 diabetes. Diabetes 53(12):3048–3056

Larsen RN, Mann NJ, Maclean E, Shaw JE (2011) The effect of high-protein, low-carbohydrate diets in the treatment of type 2 diabetes: a 12 month randomised controlled trial. Diabetologia 54(4):731–740. doi:10.1007/s00125-010-2027-y

Levitan EB, Song Y, Ford ES, Liu S (2004) Is nondiabetic hyperglycemia a risk factor for cardiovascular disease? A meta-analysis of prospective studies. Arch Intern Med 164(19):2147–2155. doi:10.1001/archinte.164.19.2147

Liao Z, Li Y, Yao B, Fan H, Hu G, Weng J (2001) The effects of isomaltulose on blood glucose and lipids for diabetic subjects. Diabetes 50(Suppl 2):A366

Liese AD, Weis KE, Schulz M, Tooze JA (2009) Food intake patterns associated with incident type 2 diabetes. Diabetes Care 32(2):263–268

Lin HC, Doty JE, Reedy TJ, Meyer JH (1989) Inhibition of gastric emptying by glucose depends on length of intestine exposed to nutrient. Am J Physiol 256(2 Pt 1):G404–G411

Lina BAR, Jonker D, Kozianowski G (2002) Isomaltulose (Palatinose): a review of biological and toxicological studies. Food Chem Toxicol 40(10):1375–1381

Linn T, Geyer R, Prassek S, Laube H (1996) Effect of dietary protein intake on insulin secretion and glucose metabolism in insulin-dependent diabetes mellitus. J Clin Endocrinol Metab 81(11):3938–3943

Linn T, Santosa B, Gronemeyer D, Aygen S, Scholz N, Busch M, Bretzel RG (2000) Effect of long-term dietary protein intake on glucose metabolism in humans. Diabetologia 43(10):1257–1265

Little TJ, Doran S, Meyer JH, Smout AJPM, O'Donovan DG, Wu KL, Jones KL, Wishart J, Rayner CK, Horowitz M, Feinle-Bisset C (2006) The release of GLP-1 and ghrelin, but not GIP and CCK, by glucose is dependent upon the length of small intestine exposed. Am J Physiol Endocrinol Metab 291(3):E647–E655. doi:10.1152/ajpendo.00099.2006

Low NH, Sporns P (1988) Analysis and quantitation of minor di- and trisaccharides in honey, using capillary gas chromatography. J Food Sci 53(2):558–561

Ludvik B, Nolan JJ, Roberts A, Baloga J, Joyce M, Bell JM, Olefsky JM (1997) Evidence for decreased splanchnic glucose uptake after oral glucose administration in non-insulin-dependent diabetes mellitus. J Clin Invest 100(9):2354–2361. doi:10.1172/JCI119775

Macdonald I, Daniel J (1983) The bio-availability of isomaltulose in man and rat. Nutr Rep Int 28(5):1083–1090

Maeda A, Miyagawa Ji, Miuchi M, Nagai E, Konishi K, Matsuo T, Tokuda M, Kusunoki Y, Ochi H, Murai K, Katsuno T, Hamaguchi T, Harano Y, Namba M (2013) Effects of the naturally-occurring disaccharides, palatinose and sucrose, on incretin secretion in healthy non-obese subjects. J Diabetes Investig 4(3):281–286

Magnusson I, Rothman DL, Katz LD, Shulman RG, Shulman GI (1992) Increased rate of gluconeogenesis in type II diabetes mellitus. A 13C nuclear magnetic resonance study. J Clin Invest 90(4):1323–1327. doi:10.1172/JCI115997

Magnusson I, Rothman DL, Gerard DP, Katz LD, Shulman GI (1995) Contribution of hepatic glycogenolysis to glucose production in humans in response to a physiological increase in plasma glucagon concentration. Diabetes 44(2):185–189

Manders RJ, Wagenmakers AJ, Koopman R, Zorenc AH, Menheere PP, Schaper NC, Saris WH, van Loon LJ (2005) Co-ingestion of a protein hydrolysate and amino acid mixture with carbohydrate improves plasma glucose disposal in patients with type 2 diabetes. Am J Clin Nutr 82(1):76–83

Manders RJ, Koopman R, Sluijsmans WE, Van Den Berg R, Verbeek K, Saris WH, Wagenmakers AJ, Van Loon LJ (2006) Co-ingestion of a protein hydrolysate with or without additional leucine effectively reduces postprandial blood glucose excursions in Type 2 diabetic men. J Nutr 136(5):1294–1299

Mann JI, De Leeuw I, Hermansen K, Karamanos B, Karlström B, Katsilambros N, Riccardi G, Rivellese AA, Rizkalla S, Slama G, Toeller M, Uusitupa M, Vessby B (2004) Evidence-based nutritional approaches to the treatment and prevention of diabetes mellitus. Nutr Metab Cardiovasc Dis 14(6):373–394

Mari A (1992) Estimation of the rate of appearance in the non-steady state with a two-compartment model. Am J Physiol 263(2 Pt 1):E400–E415

Matthews DR, Hosker JP, Rudenski AS, Naylor BA, Treacher DF, Turner RC (1985) Homeostasis model assessment: insulin resistance and beta-cell function from fasting plasma glucose and insulin concentrations in man. Diabetologia 28(7):412–419

Menzies IS (1974) Absorption of intact oligosaccharide in health and disease. Biochem Soc Trans 2:1042–1047

Meyer C, Dostou J, Nadkarni V, Gerich J (1998a) Effects of physiological hyperinsulinemia on systemic, renal, and hepatic substrate metabolism. Am J Physiol 275(6 Pt 2):F915–F921

Meyer JH, Tabrizi Y, DiMaso N, Hlinka M, Raybould HE (1998b) Length of intestinal contact on nutrient-driven satiety. Am J Physiol 275(4 Pt 2):R1308–R1319

Meyer C, Saar P, Soydan N, Eckhard M, Bretzel RG, Gerich J, Linn T (2005) A potential important role of skeletal muscle in human counterregulation of hypoglycemia. J Clin Endocrinol Metab 90(11):6244–6250

Milicevic Z, Raz I, Beattie SD, Campaigne BN, Sarwat S, Gromniak E, Kowalska I, Galic E, Tan M, Hanefeld M (2008) Natural history of cardiovascular disease in patients with diabetes: role of hyperglycemia. Diabetes Care 31(Suppl 2):S155–S160. doi:10.2337/dc08-s240

Millipore (2011) Glucagon RIA Kit 250 tubes (Cat. # GL-32K)

Millipore (2012a) Glucagon-like peptide-1 (active) ELISA kit 96-well plate (Cat. # EGLP-35K)

Millipore (2012b) Human GIP (total) ELISA kit 96-well plate (Cat. # EZHGIP-54K)

Mitrakou A, Kelley D, Veneman T, Jenssen T, Pangburn T, Reilly J, Gerich J (1990) Contribution of abnormal muscle and liver glucose metabolism to postprandial hyperglycemia in NIDDM. Diabetes 39(11):1381–1390

Muniyappa R, Lee S, Chen H, Quon MJ (2008) Current approaches for assessing insulin sensitivity and resistance in vivo: advantages, limitations, and appropriate usage. Am J Physiol Endocrinol Metab 294(1):E15–E26. doi:10.1152/ajpendo.00645.2007

Nathan DM, Buse JB, Davidson MB, Ferrannini E, Holman RR, Sherwin R, Zinman B (2009) Medical management of hyperglycemia in type 2 diabetes: a consensus algorithm for the initiation and adjustment of therapy: a consensus statement of the American Diabetes Association and the European Association for the Study of Diabetes. Diabetes Care 32(1):193–203. doi:10.2337/dc08-9025

Nauck MA, Homberger E, Siegel EG, Allen RC, Eaton RP, Ebert R, Creutzfeldt W (1986) Incretin effects of increasing glucose loads in man calculated from venous insulin and C-peptide responses. J Clin Endocrinol Metab 63(2):492–498

Nauck MA, Heimesaat MM, Orskov C, Holst JJ, Ebert R, Creutzfeldt W (1993) Preserved incretin activity of glucagon-like peptide 1 [7-36 amide] but not of synthetic human gastric inhibitory polypeptide in patients with type-2 diabetes mellitus. J Clin Invest 91(1):301–307. doi:10.1172/JCI116186

Nilsson M, Stenberg M, Frid AH, Holst JJ, Bjorck IM (2004) Glycemia and insulinemia in healthy subjects after lactose-equivalent meals of milk and other food proteins: the role of plasma amino acids and incretins. Am J Clin Nutr 80(5):1246–1253

O'Donovan DG, Doran S, Feinle-Bisset C, Jones KL, Meyer JH, Wishart JM, Morris HA, Horowitz M (2004) Effect of variations in small intestinal glucose delivery on plasma glucose, insulin, and incretin hormones in healthy subjects and type 2 diabetes. J Clin Endocrinol Metab 89(7):3431–3435. doi:10.1210/jc.2004-0334

Patterson BW (1997) Use of stable isotopically labeled tracers for studies of metabolic kinetics: an overview. Metabolism 46(3):322–329

Porter MC, Kuijpers MH, Mercer GD, Hartnagel RE, Koeter HB (1991) Safety evaluation of Protaminobacter rubrum: intravenous pathogenicity and toxigenicity study in rabbits and mice. Food Chem Toxicol 29(10):685–688

Pounis GD, Tyrovolas S, Antonopoulou M, Zeimbekis A, Anastasiou F, Bountztiouka V, Metallinos G, Gotsis E, Lioliou E, Polychronopoulos E, Lionis C, Panagiotakos DB (2010) Long-term animal-protein consumption is associated with an increased prevalence of diabetes among the elderly: the Mediterranean Islands (MEDIS) study. Diabetes Metab 36(6 Pt 1): 484–490

Psaltopoulou T, Ilias I, Alevizaki M (2010) The role of diet and lifestyle in primary, secondary, and tertiary diabetes prevention: a review of meta-analyses. Rev Diabet Stud 7(1):26–35

Qualmann C, Nauck MA, Holst JJ, Orskov C, Creutzfeldt W (1995) Glucagon-like peptide 1 (7-36 amide) secretion in response to luminal sucrose from the upper and lower gut. A study using alpha-glucosidase inhibition (acarbose). Scand J Gastroenterol 30(9):892–896

Radziuk J, Norwich KH, Vranic M (1978) Experimental validation of measurements of glucose turnover in nonsteady state. Am J Physiol 234(1):E84–E93

Reinehr T (2013) Type 2 diabetes mellitus in children and adolescents. World J Diabetes 4(6):270–281. doi:10.4239/wjd.v4.i6.270

Rennie MJ (1999) An introduction to the use of tracers in nutrition and metabolism. Proc Nutr Soc 58(4):935–944

Ripsin CM, Kang H, Urban RJ (2009) Management of blood glucose in type 2 diabetes mellitus. Am Fam Physician 79(1):29–36

Rizkalla SW, Taghrid L, Laromiguiere M, Huet D, Boillot J, Rigoir A, Elgrably F, Slama G (2004) Improved plasma glucose control, whole-body glucose utilization, and lipid profile on a low-glycemic index diet in type 2 diabetic men: a randomized controlled trial. Diabetes Care 27(8):1866–1872

Rizza RA, Mandarino LJ, Gerich JE (1981) Dose-response characteristics for effects of insulin on production and utilization of glucose in man. Am J Physiol 240(6):E630–E639

Sakurai Y, Ochiai M, Funabiki T (2000) Assessment of in vivo glucose kinetics using stable isotope tracers to determine their alteration in humans during critical illness. Surg Today 30(1):1–10

Schirra J, Katschinski M, Weidmann C, Schäfer T, Wank U, Arnold R, Göke B (1996) Gastric emptying and release of incretin hormones after glucose ingestion in humans. J Clin Invest 97(1):92–103. doi:10.1172/JCI118411

Seifarth C, Bergmann J, Holst JJ, Ritzel R, Schmiegel W, Nauck MA (1998) Prolonged and enhanced secretion of glucagon-like peptide 1 (7-36 amide) after oral sucrose due to alpha-glucosidase inhibition (acarbose) in Type 2 diabetic patients. Diabet Med 15(6):485–491. doi:10.1002/(SICI)1096-9136(199806)15:6<485::AID-DIA610>3.0.CO;2-Y

Sherwin RS, Kramer KJ, Tobin JD, Insel PA, Liljenquist JE, Berman M, Andres R (1974) A model of the kinetics of insulin in man. J Clin Invest 53(5):1481–1492. doi:10.1172/JCI107697

Siddiqui I, Furgala B (1967) Isolation and characterization of oligosaccharides from honey. Part I. Disaccharides. J Apic Res 6:139–145

Siesjö BK (1988) Hypoglycemia, brain metabolism, and brain damage. Diabetes Metab Rev 4(2):113–144

Simrén M, Stotzer PO (2006) Use and abuse of hydrogen breath tests. Gut 55(3):297–303. doi:10.1136/gut.2005.075127

Singhal P, Caumo A, Carey PE, Cobelli C, Taylor R (2002) Regulation of endogenous glucose production after a mixed meal in type 2 diabetes. Am J Physiol Endocrinol Metab 283(2):E275–E283. doi:10.1152/ajpendo.00424.2001

Slein MW (1965) D-glucose: determination with hexokinase and glucose-6-phosphate dehydrogenase. In: Bergmeyer HU (ed) Methods of enzymatic analysis (second printing, revised edn.). Academic Press, pp 117–130. http://dx.doi.org/10.1016/B978-0-12-395630-9.50023-2

Slein MW, Cori GT, Cori CF (1950) A comparative study of hexokinase from yeast and animal tissues. J Biol Chem 186(2):763–780

Sluijs I, Beulens JWJ, Van Der A DL, Spijkerman AMW, Grobbee DE, Van Der Schouw YT (2010) Dietary intake of total, animal, and vegetable protein and risk of type 2 diabetes in the European Prospective Investigation into Cancer and Nutrition (EPIC)-NL study. Diabetes Care 33(1):43–48

Solomon TP, Haus JM, Kelly KR, Cook MD, Filion J, Rocco M, Kashyap SR, Watanabe RM, Barkoukis H, Kirwan JP (2010) A low-glycemic index diet combined with exercise reduces insulin resistance, postprandial hyperinsulinemia, and glucose-dependent insulinotropic polypeptide responses in obese, prediabetic humans. Am J Clin Nutr 92(6):1359–1368. doi: 10.3945/ajcn.2010.29771

Soop M, Nygren J, Brismar K, Thorell A, Ljungqvist O (2000) The hyperinsulinaemic-euglycaemic glucose clamp: reproducibility and metabolic effects of prolonged insulin infusion in healthy subjects. Clin Sci 98(4):367–374

Steele R (1959) Influences of glucose loading and of injected insulin on hepatic glucose output. Ann N Y Acad Sci 82:420–430

Steele R, Wall JS, De Bodo RC, Altszuler N (1956) Measurement of size and turnover rate of body glucose pool by the isotope dilution method. Am J Physiol 187(1):15–24

Steele R, Bjerknes C, Rathgeb I, Altszuler N (1968) Glucose uptake and production during the oral glucose tolerance test. Diabetes 17(7):415–421

Steele R, Rostami H, Altszuler N (1974) A two-compartment calculator for the dog glucose pool in the nonsteady state. Fed Proc 33(7):1869–1876

Stegink LD, Filer Jr LJ, Baker GL (1983) Effect of carbohydrate on plasma and erythrocyte glutamate levels in humans ingesting large doses of monosodium L-glutamate in water. Am J Clin Nutr 37(6):961–968

Stumvoll M, Perriello G, Nurjhan N, Bucci A, Welle S, Jansson PA, Dailey G, Bier D, Jenssen T, Gerich J (1996) Glutamine and alanine metabolism in NIDDM. Diabetes 45(7):863–868

Sugar Nutrition UK (2011) Worldwide sugar production and consumption data. http://www.sugarnutrition.org.uk/Worldwide-sugar-production-and-consumption-data.aspx. Last visited 31 Mar 2014

Takazoe I (1985) New trends on sweeteners in Japan. Int Dent J 35(1):58–65

Tamura A, Shiomi T, Tamaki N, Shigematsu N, Tomita F, Hara H (2004) Comparative effect of repeated ingestion of difructose anhydride III and palatinose on the induction of gastrointestinal symptoms in humans. Biosci Biotechnol Biochem 68(9):1882–1887

Tang JE, Moore DR, Kujbida GW, Tarnopolsky MA, Phillips SM (2009) Ingestion of whey hydrolysate, casein, or soy protein isolate: effects on mixed muscle protein synthesis at rest and following resistance exercise in young men. J Appl Physiol 107(3):987–992

Thomas DE, Elliott EJ (2010) The use of low-glycaemic index diets in diabetes control. Br J Nutr 104(6):797–802

Thomas GN, Jiang CQ, Taheri S, Xiao ZH, Tomlinson B, Cheung BMY, Lam TH, Barnett AH, Cheng KK (2010) A systematic review of lifestyle modification and glucose intolerance in the prevention of type 2 diabetes. Curr Diabetes Rev 6(6):378–387

Tinker LF, Sarto GE, Howard BV, Huang Y, Neuhouser ML, Mossavar-Rahmani Y, Beasley JM, Margolis KL, Eaton CB, Phillips LS, Prentice RL (2011) Biomarker-calibrated dietary energy and protein intake associations with diabetes risk among postmenopausal women from the Women's Health Initiative. Am J Clin Nutr 94(6):1600–1606. doi:10.3945/ajcn.111.018648

Tonouchi H, Yamaji T, Uchida M, Koganei M, Sasayama A, Kaneko T, Urita Y, Okuno M, Suzuki K, Kashimura J, Sasaki H (2011) Studies on absorption and metabolism of palatinose (isomaltulose) in rats. Br J Nutr 105(1):10–14. doi:10.1017/S0007114510003193

Triadou N, Bataille J, Schmitz J (1983) Longitudinal study of the human intestinal brush border membrane proteins. Distribution of the main disaccharidases and peptidases. Gastroenterology 85(6):1326–1332

Triplitt CL (2012a) Examining the mechanisms of glucose regulation. Am J Manag Care 18(1 Suppl):S4–S10

Triplitt CL (2012b) Understanding the kidneys' role in blood glucose regulation. Am J Manag Care 18(1 Suppl):S11–S16

Tsuji Y, Yamada K, Hosoya N, Moriuchi S (1986) Digestion and absorption of sugars and sugar substitutes in rat small intestine. J Nutr Sci Vitaminol 32(1):93–100

Turner RC, Holman RR, Matthews D, Hockaday TD, Peto J (1979) Insulin deficiency and insulin resistance interaction in diabetes: estimation of their relative contribution by feedback analysis from basal plasma insulin and glucose concentrations. Metabolism 28(11):1086–1096

Turner RC, Mathews DR, Holman RR, Peto J (1982) Relative contributions of insulin deficiency and insulin resistance in maturity-onset diabetes. Lancet 1(8272):596–598

van Can JGP, Ijzerman TH, van Loon LJC, Brouns F, Blaak EE (2009) Reduced glycaemic and insulinaemic responses following isomaltulose ingestion: implications for postprandial substrate use. Br J Nutr 102(10):1408–1413. doi:10.1017/S0007114509990687

van Can JGP, van Loon LJC, Brouns F, Blaak EE (2012) Reduced glycaemic and insulinaemic responses following trehalose and isomaltulose ingestion: implications for postprandial substrate use in impaired glucose-tolerant subjects. Br J Nutr 108(7):1210–1217. doi:10.1017/S0007114511006714

van Loon LJ, Kruijshoop M, Menheere PP, Wagenmakers AJ, Saris WH, Keizer HA (2003) Amino acid ingestion strongly enhances insulin secretion in patients with long-term type 2 diabetes. Diabetes Care 26(3):625–630

Veldhorst MA, Nieuwenhuizen AG, Hochstenbach-Waelen A, van Vught AJ, Westerterp KR, Engelen MP, Brummer RJ, Deutz NE, Westerterp-Plantenga MS (2009) Dose-dependent satiating effect of whey relative to casein or soy. Physiol Behav 96(4–5):675–682

Vella A, Rizza RA (2009) Application of isotopic techniques using constant specific activity or enrichment to the study of carbohydrate metabolism. Diabetes 58(10):2168–2174. doi:10.2337/db09-0318

Vella A, Shah P, Basu R, Basu A, Holst JJ, Rizza RA (2000) Effect of glucagon-like peptide 1(7-36) amide on glucose effectiveness and insulin action in people with type 2 diabetes. Diabetes 49(4):611–617

Vilsbøll T, Krarup T, Deacon CF, Madsbad S, Holst JJ (2001) Reduced postprandial concentrations of intact biologically active glucagon-like peptide 1 in type 2 diabetic patients. Diabetes 50(3):609–613

Wahren J, Ekberg K (2007) Splanchnic regulation of glucose production. Annu Rev Nutr 27:329–345. doi:10.1146/annurev.nutr.27.061406.093806

Walker KZ, O'Dea K, Gomez M, Girgis S, Colagiuri R (2010) Diet and lifestyle in the prevention of the rising diabetes pandemic. J Hum Nutr Diet 23(4):333–335

Wang ET, de Koning L, Kanaya AM (2010) Higher protein intake is associated with diabetes risk in South Asian Indians: the Metabolic Syndrome and Atherosclerosis in South Asians Living in America (MASALA) study. J Am Coll Nutr 29(2):130–135

West DJ, Morton RD, Stephens JW, Bain SC, Kilduff LP, Luzio S, Still R, Bracken RM (2011) Isomaltulose improves postexercise glycemia by reducing CHO oxidation in T1DM. Med Sci Sports Exerc 43(2):204–210. doi:10.1249/MSS.0b013e3181eb6147

WHO (1988) Toxicological evaluation of certain food additives. Joint FAO/WHO Expert Committee on Food Additives (JECFA). WHO Food Additives Series No. 22. Geneva

WHO (1999) Definition, diagnosis and classification of diabetes mellitus and its complications: report of a WHO consultation. Part 1: diagnosis and classification of diabetes mellitus. Technical report, World Health Organization, Geneva

WHO (2006) Definition and diagnosis of diabetes mellitus and intermediate hyperglycaemia: report of a WHO/IDF consultation. Technical report, World Health Organization, Geneva

WHO (2011a) Global status report on noncommunicable diseases 2010. Technical report, World Health Organization, Geneva

WHO (2011b) Use of glycated haemoglobin (HbA1c) in the diagnosis of diabetes mellitus: abbreviated report of a WHO consultation. Technical report, World Health Organization, Geneva

Wiecko J, Sherman WR (1976) Boroacetylation of carbohydrates. Correlations between structure and mass spectral behavior in monoacetylhexose cyclic boronic esters. J Am Chem Soc 98(24):7631–7637

Willms B, Werner J, Holst JJ, Orskov C, Creutzfeldt W, Nauck MA (1996) Gastric emptying, glucose responses, and insulin secretion after a liquid test meal: effects of exogenous glucagon-like peptide-1 (GLP-1)-(7-36) amide in type 2 (noninsulin-dependent) diabetic patients. J Clin Endocrinol Metab 81(1):327–332. doi:10.1210/jcem.81.1.8550773

Woerle HJ, Meyer C, Dostou JM, Gosmanov NR, Islam N, Popa E, Wittlin SD, Welle SL, Gerich JE (2003) Pathways for glucose disposal after meal ingestion in humans. Am J Physiol Endocrinol Metab 284(4):E716–E725. doi:10.1152/ajpendo.00365.2002

Wolever TMS, Mehling C (2002) High-carbohydrate-low-glycaemic index dietary advice improves glucose disposition index in subjects with impaired glucose tolerance. Br J Nutr 87(5):477–487. doi:10.1079/BJNBJN2002568

Zeleznik AJ, Roth J (1978) Demonstration of the insulin receptor in vivo in rabbits and its possible role as a reservoir for the plasma hormone. J Clin Invest 61(5):1363–1374. doi:10.1172/JCI109054

Index

M. Ang, *Metabolic Response of Slowly Absorbed Carbohydrates in Type 2 Diabetes Mellitus*, SpringerBriefs in Systems Biology, DOI 10.1007/978-3-319-27898-8

Printed in the United States
By Bookmasters